The Patrick Moore Practical Astronomy Series

More information about this series at http://www.springer.com/series/3192

Video Astronomy on the Go

Using Video Cameras
With Small Telescopes

Joseph Ashley

Joseph Ashley
Marathon, Greece

ISSN 1431-9756   ISSN 2197-6562 (electronic)
The Patrick Moore Practical Astronomy Series
ISBN 978-3-319-46935-5   ISBN 978-3-319-46937-9 (eBook)
DOI 10.1007/978-3-319-46937-9

Library of Congress Control Number: 2016954312

© Springer International Publishing AG 2017
This work is subject to copyright. All rights are reserved by the Publisher, whether the whole or part of the material is concerned, specifically the rights of translation, reprinting, reuse of illustrations, recitation, broadcasting, reproduction on microfilms or in any other physical way, and transmission or information storage and retrieval, electronic adaptation, computer software, or by similar or dissimilar methodology now known or hereafter developed.
The use of general descriptive names, registered names, trademarks, service marks, etc. in this publication does not imply, even in the absence of a specific statement, that such names are exempt from the relevant protective laws and regulations and therefore free for general use.
The publisher, the authors and the editors are safe to assume that the advice and information in this book are believed to be true and accurate at the date of publication. Neither the publisher nor the authors or the editors give a warranty, express or implied, with respect to the material contained herein or for any errors or omissions that may have been made.

Printed on acid-free paper

This Springer imprint is published by Springer Nature
The registered company is Springer International Publishing AG
The registered company address is: Gewerbestrasse 11, 6330 Cham, Switzerland

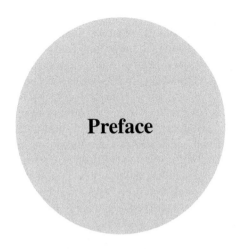

Preface

Welcome to the New World of Video Astronomy!

Today video astronomy is leaving the experimental world and entering into the mainstream of amateur astronomy. There are many types of video devices used in amateur astronomy: web cameras, planetary imagers, and even DSLRs to name a few. However, none of these have excited amateur astronomers quite like video produced by low-light closed circuit television (CCTV) security cameras which can image dim nebulae or the bright Sun and do so in color and in real time.

Video Astronomy on the Go provides an orientation into the world of video astronomy. It is not a cookbook but an overview of the technology based upon relatively inexpensive, entry-level equipment. It is intended for people with little to no knowledge of video astronomy but who have a basic working knowledge of traditional visual astronomy. The book's objective is to sufficiently discuss the various aspects of video astronomy so that a beginner can reasonably have the confidence needed to assemble and use an astro-video camera with a telescope.

Just what is video astronomy? This question can often create intense debates on the Internet between advocates of different types of cameras. Web cameras, planetary imagers, DSLRs, etc. produce a video output. However, none have the wide variety of applications associated with low-light or star light CCTV security cameras.

For the past 15–20 years or so a growing group of dedicated amateur astronomers experimented with viewing the night sky with their telescopes using a variety of video cameras and associated technologies. Some of the more obvious are web cameras, DSLR movies, video recorders, Internet Protocol (IP) security cameras, planetary imagers, and closed circuit TV (CCTV) security cameras.

Initially video technology was suited only for the moon and some planets as deep space applications were hampered by the inadequate low-light sensitivity of the available cameras. In the middle of the first decade of the twenty-first century, CCTV surveillance cameras with excellent low-light sensitivities appeared on the market. These cameras were designed for no light situations where infrared painters were unacceptable for security reasons yet a high-quality image was needed. Astronomer experimenters immediately recognized the potentials of the low-light CCTV security cameras. Soon they were using them for video telescopes to view the night sky, broadcast live images over the Internet, or photograph deep space objects. The "eye" of the camera produced real-time images far deeper in space than visible in an eyepiece and could also produce the image in color. The low-light CCTV cameras produced images previously seen only in telescopes having apertures two to three times greater than the telescope used with the video camera and also restored detail and contrast to objects washed out by artificial sky glow.

Earlier experimenters purchased and personally altered their own CCTV cameras. This limited participation to people having the skills and fortitude to adapt their own camera. As the technology matured, the demand for CCTV cameras already adapted for astronomy increased. Currently, a few small, enterprising companies not only modify the better CCTV cameras but also add additional features and capabilities to them. These modified cameras are very capable, easy to use, and readily available on the commercial market. Prices start at $100 in the United States and range upwards to $2000. The introduction of commercially available CCTV cameras ready for astronomy usage was a paradigm shift and moved video astronomy from the world of the tinkerer to the mainstream of amateur astronomy.

These modified CCTV cameras are ready to use out of the box. There is no need for the astronomer to tinker with or modify one. No computer is required as they have an analog video output and can be directly connected to a television set, monitor, or DVD player. Many terms are used to identify these modified CCTV low-light, security cameras. You will see or hear them referred to as an astro-video camera, astro video camera, astrovideo camera, astronomy camera, or an astronomical camera. The term "astro-video camera" is used in this book. This choice is purely arbitrary.

Back to the question, "Just what is video astronomy?" Essentially, video astronomy is nothing more than replacing the eyepiece of a telescope with a small, lightweight television camera. At the present time the prevailing definition and the one used in this book is video produced by an astro-video camera (a low-light CCTV camera altered for astronomy that has an analog video output signal). The most basic system has a television set connected directly to the camera. To observe with this "video telescope" one simply looks at the television screen. More complex configurations exist which allow digitizing the analog signal from a video camera and feeding the signal to a computer.

What is the advantage of a video telescope over the simplicity of using an eyepiece? This is a valid question. The telescope is outside with the cold of winter and insects of summer. The television monitor can be inside and the telescope is easily remotely controlled. The observer can sit inside in comfort and tour the night sky. By having the image on a television screen, several people can see the view in the telescope at the same time. This is ideal for public outreach programs or for mom, dad, and the kids viewing the cosmos together. The output from the television camera is easily converted into a digital signal and amateur astronomers can use the Internet to share the view in their telescopes with others around the globe. Initially, these attributes (viewing in comfort and sharing images) were two primary drivers for using video technology. Later we will discuss the third driver—astrophotography.

As CCTV camera technology advanced, the importance of viewing in comfort and the sharing of images was surpassed by the "ability to view." Some CCTV cameras became so sensitive that objects too faint to see through an eyepiece were easily visible on the television screen. This essentially meant that the effective aperture of a telescope was doubled to tripled. The view through a 102 mm video telescope rivaled the view in the eyepiece of a 200–300 mm aperture telescope. But there was an added bonus. Not only could previously invisible objects be seen, the new advances in camera technology provided the views in color.

Recently some enterprising amateurs connected a 3D converter and a 3D television set to the output of their video telescopes. The few people who have viewed through a 3D video telescope report the views of nebulae are almost like a religious experience. This aspect of video astronomy is in its infancy but has a bright promise for the future.

Earlier mention was made that astrophotography was also one of the original reasons why amateur astronomers experimented with video cameras. Here too, advances in camera technology are to the point where small inexpensive television cameras can produce images of a wide variety of deep space objects that rival many images made using dedicated astronomical CCD cameras or digital single lens reflex cameras.

As amateurs experimented using astro-video cameras for astrophotography, another interesting attribute became readily apparent. With the proper camera settings, the artificial sky glow of large metropolitan areas no longer drowned out objects in deep space. Previously invisible nebulae and galaxies became visible and faintly visible objects now had details. While nothing beats a dark sky, the video telescope provides a tool to help mitigate the impact of the artificial sky glow so prevalent in our modern society.

So, What Is *Video Astronomy on the Go* All About?

Video Astronomy on the Go is about using cameras that are available on the commercial astronomy market as "branded" astro-video cameras. These cameras are ready to use out of the box. There is no need for the astronomer to tinker with or modify a camera. They have an analog video output and can be directly connected to a television set, TV monitor, or DVD player. This is not to say that the book is not usable by someone who buys and modifies a CCTV security camera or any other television camera, only that the mechanics of the needed modifications are not discussed.

The term "astronomical CCD" camera is used to define the typical CCD camera used by astrophotographers to take long exposure images of deep space objects.

Video Astronomy on the Go is especially about using astro-video cameras with small lightweight telescopes that are easily transported by foot, on public transportation, or private automobile.

Video Astronomy on the Go is intended for amateur astronomers of all skill and knowledge levels who wish to enter the world of video astronomy. The major aspects of video astronomy are explored using an inexpensive, commercially available, astro-video camera with an entry level, goto alt-azimuth telescope so a newcomer to astronomy will understand the capabilities of a basic system. As with many aspects of amateur astronomy, the equipment used for video astronomy can be basic or advanced, simple or complex, moderately priced, or astronomically expensive. One can view from their backyard or from an observatory. One interesting aspect of video astronomy is that the capabilities of basic, moderately priced, systems can compete with the more advanced and costly systems over a wide range of activities. The major differences between the two are how deep in space one can go and the amount of noise in the images produced.

Video Astronomy on the Go is an excellent book for existing amateur astronomers and beginners who live in large metropolitan or urban areas with significant levels of artificial sky glow. The ability of a video telescope to penetrate artificial sky glow presents a paradigm shift for urban observing. The downtown city dweller now has the tools needed to see deep space with a kit that is easily stored in a small city apartment and transported on city buses and subways. This is true for both experienced astronomers and beginners to astronomy as well.

No longer is video astronomy the world of the tinkerer and the experimenter. It has matured into a powerful tool available for amateur astronomers to view and study the night sky. Video astronomy is now "prime time." Other than a few forums and discussion groups, few sources of information exist to help the amateur seeking to enter the world of video astronomy. This book, *Video Astronomy on the Go*, is meant to fill the current information void.

Video Astronomy on the Go had a rocky road to travel along the way to its birth. With the encouragement of John Watson of Springer and Sophia, my wife for some 50+ years, I was able to struggle through government collapses, a flash flood through our home, bank failures with their long lines in the heat of the summer just to get a few Euros from a teller machine, and a mass migration from the middle east into Europe. Thanks John. Thanks Sophia. We do live in interesting times.

Marathon, Greece   Joseph Ashley

Contents

1	**Astronomy from a Video Perspective**	1
	What Is an Astro-Video Camera?	1
	How Does Video Astronomy Work?	2
	Telescope Mounts	3
	Video Astronomy and Celestial Observing, or the True Armchair Astronomer	4
	Video Astronomy and Outreach	5
	Video Astronomy and Astrophotography	6
	Video Astronomy and Broadcasting	7
	Video Astronomy and Other Applications	7
2	**The Anatomy of a Video Camera**	9
	How Astro-Video Cameras Work	9
	Stacking	11
	Features Found on a Typical Entry-Level Astro-Video Camera	14
	Field of View	18
	Image Brightness	22
	Astro-Video Camera Settings and Adjustments	24
	AVSYSTEM Menu Options	29
	Color Menu Options	31
	Day and Night Menu Settings	31
	Effect Menu Settings	32
	Motion Menu Settings	33
	Test Bars Menu Settings	33
	Procamp Menu Settings	34

	SYSTEM MENU settings	34
	Exit Menu Settings	35
	How Does a Video Camera Work?	35
	Integrating Cameras	35
	Astro-Video Camera Screen Refresh	37
	General Notes Regarding Entry-Level Astro-Video Cameras	37
	A Sample of Currently Available Entry-Level Astro-Video Cameras	38
	Astro Video Systems	38
	MallinCam	39
	Revolution Imager	39
	Lntech	39
3	**Assembling Your Video Astronomy Kit**	**41**
	Getting Started	41
	Astro-Video Cameras	42
	Telescopes and Their Impact	46
	Telescope Mounts	48
	Budget Entries into Video Astronomy	52
4	**Light Pollution and Filters**	**55**
	What Causes Light Pollution?	55
	Trespass Light and How to Mitigate It	57
	Natural Sky Glow	60
	The Nature of Artificial Sky Glow	61
	How Astro-Video Cameras Pierce Artificial Sky Glow	63
	Light Pollution Reduction Filters and Astro-Video Cameras	67
	Urban Viewing in Areas with Significant Light Pollution	71
5	**The Solar System and Video Astronomy**	**73**
	Our Solar System in Brief	73
	The Advantages of Video Telescopes for Viewing the Sun	75
	Video Telescope Attributes for Exploring the Moon	80
	Lunar Observation	80
	Lunar Photography	81
	Lunar Impact Monitoring Program	82
	Occultations	83
	Video Telescope Attributes for Viewing the Planets	84
	Astro-Video Camera Attributes for Photographing Smaller Solar System Objects	86
6	**Deep Space and Video Astronomy**	**89**
	Our Window into the Universe	89
	Issues Related to Deep Space Objects and Video Astronomy	92
	Camera-Related Issues	93
	Viewing Deep Space with a Video Telescope	95
	Maksutov Cassegrain Telescopes and Video Astronomy	105

Contents xi

7   Imaging the Night Sky 109
   The Versatility of Video Imaging 109
   Analog-to-Digital Conversion 111
   Digital Processing Overview 113
   Setting Up for Astrophotography with an Astro-Video Camera 114
   Computer Programs Used in Astro-Video Photography 118
   Astronomy Video Capture and Stacking Programs 122

8   Outreach with Video Telescopes 125
   Background 125
   Video Telescope Appropriateness 126
   Video Component Considerations 127
   Logistics 128
   Video Equipment Needed 129
   Urban Sidewalk Outreach Video Telescope 130

9   Live Video Broadcasting 133
   Equipment Needed 133
   Broadcasting Sites 133
   Video Astronomy Live Website 134
   Night Skies Network Website 136
   Astronomy Live Website 137
   Planning and Executing a Live Broadcast 138
   The Rules 139

10 Video Astronomy Trends 141
   The Tricky Business of Prediction 141
   Three-Dimensional Astronomy 142
      Current Status 142
      Converting Two Dimensional Photographs
         into Three-Dimensional Anaglyphs 142
      Converting Two Dimensional Video into Three-Dimensional
         Anaglyphs 143
      Near Real-Time Three-Dimensional Video Astronomy 143
      Real-Time Three-Dimensional Astronomy 145
      3D Video Astronomy Wrap-Up 146
   Windows 10 Tablets and Video Astronomy 146
      Should You Use a Tablet with a Video Telescope? 146
      Current Tablet Usage 148
      Case Study of an Inexpensive Low-End Windows Tablet 148
   Results 149
      Suggested Windows 10 Tablet Specifications 150
      Conclusions 151
   Digital Astro-Video Cameras 151
   Urban Astronomy 152

Appendix A	**Glossary of Terms**	155
Appendix B	**Maximum Exposure Time Tables Based on 0.125 Degrees of Field Rotation**	165
Appendix C	**Star Charts for Urban Areas with Significant Light Pollution**	169
Appendix D	**Rack and Pinion Focuser Tune-UP**	195
Index		199

About the Author

An American by birth, Joseph Ashley currently lives in Greece. He has a BS in Physics, an MS in Mechanical Engineering, and a Doctorate in Public Administration. His short career in the US Navy included the recovery of astronauts Conrad and Cooper and their Gemini V spaceship from the sea; from there he began an engineering and research career involving submarine noise, chemical warfare defense, and energy conservation. Now retired, he completed his career as the Program Manager for the US Department of the Navy and Marine Corps Energy Conservation Program. He purchased his first telescope, a Meade 2045LX3, in 1987 and still uses the little 102 mm SCT, along with other scopes, until this day including his favorite, a C6S (150 mm SCT) on a lightweight alt-azimuth GOTO mount. After retirement, Ashley participated in online astronomy forums, primarily the Astronomy Forum, but does drop by the Stargazers Lounge from time to time. In late 2009 he became a moderator on The Astronomy Forum. Parallel with that, he pitched into what he calls "the dark side of astronomy"—astrophotography—concentrating on getting the best possible images from simple lightweight equipment. That took him into the world of video astronomy where today he is evaluating the advantages that small portable tablets bring to video astronomy. Previously, he published *Astrophotography on the Go; Using Short Exposures on Light Mounts* in Patrick Moore's Practical Astronomy Series.

Chapter 1

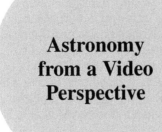

Astronomy from a Video Perspective

What Is an Astro-Video Camera?

Just what is an astro-video camera, and how does it work? Unfortunately, video astronomy is not well defined. Movie files and video clips can be made by a wide variety of cameras, including digital single lens reflex cameras. For the purposes of this book, an astro-video camera is a small television camera that displays its image using either a NTSC or PAL Television Standard analog television signal.

Video astronomy is actually nothing more than substituting the eyepiece of a telescope with an astro-video camera. To simplify matters for future discussion, we will call this combination, that of a telescope with an astro-video camera, a "video telescope." The camera provides a live, analog, video signal to a television monitor, DVD player, etc., that can be shared by more than one person. More complex configurations involving analog to digital converters, computers, Internet access, etc., are also used dependent upon exactly what the astronomer wants to accomplish.

A varied selection of astro-video cameras are on today's market. This book focuses upon entry-level cameras that typically cost about $200 with accessories (cables, remote camera controls, etc.). This does not mean that the information in it is not applicable to other astro-video cameras that have higher performance levels, only that the examples in the book as well as the other topics covered in the book are derived around the entry-level camera.

One major difference between viewing through the eyepiece with the human eye and viewing on a monitor connected to a camera is that the camera can integrate a signal over time while the human eye cannot. This ability of a camera to integrate over time has a major impact. It allows a camera to capture objects far

© Springer International Publishing AG 2017
J. Ashley, *Video Astronomy on the Go*, The Patrick Moore
Practical Astronomy Series, DOI 10.1007/978-3-319-46937-9_1

fainter than the human eye can detect. This is a well known attribute and is why astrophotographs are far more detailed than any view seen with the naked eye.

However, video astronomy is more than just making an image, as is done with an astronomical CCD camera or a digital single lens reflex camera. These images are static and typically require an extensive amount of image processing. Video astronomy is different. Like viewing through an eyepiece with your eye, video astronomy is live (real time). These two aspects, the ability to integrate over time and the ability to view real time, allows video astronomy to perform functions not possible with the human eye through a telescope's eyepiece. Actually, viewing with an astro-video camera is near real time as there can be a lag of a few milliseconds to a couple of minutes dependent upon how the camera is used.

Just what can video astronomy do that makes it worthwhile using? Here is a list giving a few applications for video astronomy. More exist and are also discussed in this book:

- penetrate light pollution
- see deeper into space
- see deep space in color
- share images real time with others
- remote viewing
- astrophotography.

How Does Video Astronomy Work?

A video telescope is a typical optical tube assembly (OTA) often adapted for video by adjusting its focal length with a focal reducer and inserting an astro-video camera in the telescope's focuser. Unlike astrophotography using DSLRs (digital single lens reflex cameras), just about any telescope will do, including Newtonians.

The astro-video camera with its analog output makes remote locating of the TV monitor or computer screen far from the telescope and camera something that is easily done. Unlike the digital signals from web cameras and astronomical CCD (charge coupled device) cameras that are so often limited by their cabling to about 4.5 m (15 feet), the analog signal from a video camera can travel upwards of 30 m (100 feet) and much more. This provides a lot of flexibility concerning the location of the viewing and recording equipment.

The actual dimensions of astro-video camera sensors are small, while the focal lengths of popular telescopes can be quite long. The combination of a small sensor and long focal length results in a small field of view with a large image size. With the exception of short focal length telescopes, such as an 80-mm f/5 short tube refractor, the image sizes are large. This creates a problem. Finding and tracking objects in the sky using an astro-video camera and a typical telescope is rather difficult. These two issues are addressed by (1) Using a focal reducer to reduce the focal length of a telescope, and (2) Using a GOTO mount to locate and track objects in space.

Although not a hard and fast rule, for video astronomy a focal length less than around 800–1000 mm works very well for the deep space objects typically observed by amateur astronomers. A 0.5x focal reducer improves the situation considerably for telescopes having a focal length longer than 1000 mm. The Meade 0.33x focal reducer is popular for many owners of SCTs as are Celestron SCTs having the Celestron/Starizona HyperStar/FastStar setup.

Finding objects without a GOTO mount is difficult but doable if you have the skills. However, you will need a tracking capability to keep the object in view once you do find it. For most people using a GOTO mount is the best, perhaps only, option for both finding and tracking objects. As far as what kind of mount to use; unless you plan to use your video telescope for astrophotography, an alt-azimuth GOTO mount works very well, almost as well as an equatorial GOTO mount in most situations.

Summing up, a typical video telescope for viewing deep space will have the following characteristics:

- an astro-video camera
- any size aperture
- focal length of less than 800 and 1000 mm either with or without using a focal reducer
- GOTO alt-azimuth or GOTO equatorial mount.

Notes Video telescope and camera characteristics will be discussed in more detail in Chaps. 2 and 3. If you are having difficulty with terms used in this discussion you may want to check out Appendix A in this book, "Glossary of Terms."

Telescope Mounts

As mentioned earlier, a GOTO mount is essentially a necessity for most video astronomy applications. The small sensor sizes of astro-video cameras produce a rather small field of view. Any mount or tripod vibration is amplified. The implication of this is that a sturdy tripod is required. However, as discussed later, this is not necessarily the case for all video astronomy applications.

A small field of view and large image size means that star hopping is by and large not possible, especially for dim objects. Setting circles are not much better. Locating and tracking objects is often tedious and frustrating. Furthermore, if you live in an urban area, artificial skyglow often makes star hopping essentially impossible. A GOTO telescope solves these issues.

Using a manually operated mount such as a Dobsonian or a German equatorial mount is technically possible but very difficult. You can easily do a test before buying a camera to see if such a setup is suited for you and your conditions. Go out at night with your telescope and only a 6-mm eyepiece. If you have difficulty locating and then viewing objects, you may seriously want to consider a GOTO telescope mount.

One advantage of video astronomy is that the observer can view from inside the comforts of home and remotely control the video telescope. Here again, an accurate GOTO mount, alt-azimuth or equatorial, is required.

Alt-azimuth or equatorial mount? Whenever an alt-azimuth mount is used with a camera, the impact of field rotation must be considered. In the past, for visual work, the issues associated with field rotation and an alt-azimuth mount were not significant. However, with the current integrating cameras having total integrated exposure times of several minutes, an equatorial mount undoubtedly produces the best images. Another issue related to an alt-azimuth mount is that many have difficulty with objects very near the zenith (tube strikes or tracking). An alt-azimuth mount on a wedge or a German equatorial mount, if precisely polar aligned, can counter the impact of field rotation. No constraints exist near the zenith, but you may have to deal with meridian flipping.

Video Astronomy and Celestial Observing, or the True Armchair Astronomer

One reason for the development of video astronomy over the years was the ability to view the night sky from inside the comfort of home; in other words, the ability to be the ultimate armchair observer (see Fig. 1.1). The system needed to do this is rather simple, requiring nothing more than an astro-video camera, a GOTO telescope, some long cables, and a television set. This basic setup requires that the observer venture outside into the elements of winter and the insects of summer to align the telescope mount and to focus the camera. If the distance is not too far, say around 100 feet (30 m), an extension can be added to the hand controller cable and the telescope controlled from inside. For longer distances, the GOTO mount can be controlled inside using a laptop computer and remote controls used for the camera settings and telescope focus.

Regardless of whether you set up your system to view the night sky from inside the comfort of your home or you sit outside in a comfortable chair holding a small television monitor in your hand, video astronomy will give you the same view of the night sky. You will see objects in depth and details not possible using an eyepiece with your telescope. In addition you will see the objects in color. An astro-video camera will take you two to three magnitudes deeper than your eyepiece. In effect, it is a multiplier that effectively increases the aperture of your telescope by up to a factor of three.

Like most things in astronomy, video astronomy loves a dark sky. However, if you are viewing from light-polluted skies, you will be pleasantly surprised. Add a light-pollution filter such as a CLS filter and objects invisible in the eyepiece are easily seen with details and color on your television screen. How deep can you penetrate artificial skyglow is dependent upon your equipment, but be prepared to see details in nebulae and galaxies and see them in color.

Fig. 1.1 Remote viewing

Video Astronomy and Outreach

Video astronomy is especially useful for outreach activities. Here, all that is required is an astro-video camera, a telescope, and an appropriately sized television set for the number of people expected. Video astronomy also requires a different way of thinking concerning outreach activities. No longer do the bright planets and a couple of deep space objects hold an outreach activity hostage. There is no need to have a long line of people waiting for a few seconds to gaze at the Moon or a planet. With a large screen television, many people can simultaneously see what the telescope sees. With video astronomy, an educational program incorporating a tour of the various kinds of objects in the night sky is possible in synchronization with a lecture on the objects being observed. Such activities are not practical with folks lined up to individually view one object at the time in the eyepiece of a telescope.

However, there is a dark side, so to speak, associated with using video astronomy for outreach activities. A television set produces a large amount of light. If the outreach event is at a star party, then the video setup must be either shielded from or located remotely from the traditional telescopes. Electrical power is also another issue, as a large television will need its share.

Video Astronomy and Astrophotography

Astrophotography requires converting the analog signal from the astro-video camera into a digital signal and then storing the image or video into the memory of a computer for processing. A video capture device (also known as a frame grabber) is used to convert the analog signal to a digital signal. Operation is simple; just plug the analog video cable into one end of the frame grabber, and it converts the analog signal into a digital signal formatted for an USB port. Your computer now sees a camera attached. Then, if you wish, a computer program stores and displays the signal from your camera.

Entry level astro-video cameras have shutter speeds that are selectable, from as short as 1/100,000 of a second to as long as 17 seconds for the NTSC standard or 20 seconds for the PAL standard. They also have the ability to reduce noise by internally digitally combining (stacking) up to five images. With a 17-second exposure, this means that a our entry-level NTSC camera can produce an image having a total integrated exposure time of 85 seconds, and a PAL camera can produce an image having a total integrated exposure time of 100 seconds. Keep in mind that a typical alt-azimuth mount can support exposure times of 20–30 seconds, which is long enough to cover most areas of the night sky before field rotation creates star trailing and blurring of the image. Also, as an object approaches the zenith, exposure times rapidly decrease, and only a couple of seconds are allowable near the zenith (See Appendix B). On the other hand, a precisely aligned equatorial mount can image the entire night sky.

The images made with an astro-video camera can be digitized and then stacked like any other image. Unless a focal reducer is used, the field of view with an astro-video camera is very small, implying a very large image size. Some stacking programs cannot adjust for field rotation, and their images may be blurred if an alt-azimuth mount is used. Other stacking programs adjust for field rotation but need a minimum number of stars common to all the images to successfully stack the image. For example, *DeepSkyStacker* requires eight stars common to all light frames. Due to the large image size and small field of view, the images of many objects do not contain eight common stars, and the program will not work. Field rotation is not an issue with a precisely aligned equatorial mount, and the probability of successfully stacking an image is far greater. More on imaging with a video telescope is discussed in Chap. 7.

Video Astronomy and Broadcasting

With video astronomy you can broadcast the view in your telescope to almost any place on our planet. All that is needed is a decent Internet connection. The easiest way to broadcast is to use a website such as Night Skies Network. With this site, which is very popular with astronomers, all you need do is establish a free account. When you log onto the site, it will establish communications with your computer, and then, after selecting a few options, you will be broadcasting live on the Night Skies Network. Anyone who wishes to see your broadcast only needs to go to the site's directory, find your name, and right click on it. In addition to the video, you also have two-way voice as well as Instant Messenger communications with anyone viewing your video stream.

Another variation is to set up your own personal site for broadcasting the signal. There are many sites on the Internet that host broadcasting sites; some offer free services, but most charge a monthly fee.

Video Astronomy and Other Applications

If you add a 3D signal converter to the output of an astro-video camera and feed the signal to a 3D television set, you will see a 3D image of the object you are viewing. People who have experienced 3D astronomy report that seeing nebulae in 3D is almost akin to a religious experience. 3D astronomy as well as substituting inexpensive tablets for laptop computers are two areas currently being explored by many amateurs.

Chapter 2

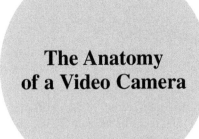

The Anatomy of a Video Camera

How Astro-Video Cameras Work

Current popular entry-level astro-video cameras as well as many advanced cameras are adaptations of closed-circuit television (CCTV) security camera technology that was designed for use in very low light situations. One such security application requires the covert production of high quality monochrome or color images or the imaging of moving targets without blurring in low light situations. This covert requirement, among other things, means that infrared painting is not allowed, as the infrared source betrays the presence of the security camera.

For this application, only the natural light from the night sky is used—approximately 0.001 lux or less for a dark sky with no Moon. These low light security cameras, or starlight cameras as they are often called, have excellent sensitivities with a minimum illumination as low as 0.0001 lux for monochrome and 0.001 lux for color. It is this starlight security camera technology that is used by current introductory level and more advanced astro-video cameras.

Video astronomy is still in its formative stage and is considered by a large segment of astronomers as wholly an amateur activity. At one time amateurs purchased CCTV security cameras and used them to explore deep space, and many still do. Today, a few very small companies, all started by amateur astronomers, produce astro-video cameras that are available on the commercial market. These cameras use the technology developed for starlight security cameras but modified to varying degrees to adapt the technology for astronomy. Recall from Chap. 1 that this book is written around the typical entry-level astro-video camera currently costing about $200 for the camera, an adapter, a video cable, a power cable, a remote controller, and a remote control cable.

Since the typical entry-level astro-video camera is based upon CCTV technology, it has an analog output that is compatible with either a NTSC or PAL Television Standard. The NTSC standard is used in North and part of South America, the Philippines, South Korea, Taiwan, and Japan, while the PAL standard, excluding the former Soviet Block countries and parts of Africa, is used in most other parts of the world. The decision of whether to have a NTSC or PAL standard camera is made at the time of purchase, as the standard does impact the physical design of the camera's sensor.

From a user's perspective, cameras using either standard operate essentially the same. With an entry-level camera, the major differences between the two standards are as follows:

- slowest shutter speed: 1/60 of a second NTSC; 1/50 of a second PAL
- longest integrated exposure time: 17 seconds NTSC; 20 seconds PAL
- longest integrated exposure time with digital noise reduction: 85 seconds NTSC; 100 seconds PAL.

Two types of sensors dominate cameras today, sensors based upon CCDs or on complementary metal oxide semiconductor (CMOS) technologies. Current astro-video cameras use a CCD sensor primarily because that is the technology used by CCTV starlight cameras. For producing color images, CCTV starlight sensors typically use a CMYC (cyan, magenta, yellow, and green) color registration. The manufacturer claims that CMYC filters are more translucent than RGB (red, green, and blue) filters, allowing more light to pass through to the pixel. The CCD sensors developed for CCTV cameras use a global shutter, meaning the entire chip is exposed at once, reducing image distortion due to movement such as a speeding automobile.

As just mentioned, the output of cameras using CCD sensors developed for starlight CCTV applications have either a PAL or NTSC analog video signal. Since neither the PAL nor the NTSC standards are high resolution, a high-resolution sensor is not needed. Although some current astronomical CCD cameras have sensors of 20 megapixels and larger, the sensor for a CCTV camera (an astro-video camera) has only around 0.5 megapixels. This has a major impact upon camera resolution.

Typical sensor sizes for CCTV cameras are 1/4, 1/3, or 1/2 of an inch with a 1/3 of an inch sensor used for entry-level astro-video cameras. Regardless of the size sensor used, the CCTV camera will typically have around 0.5 megapixels, as the camera must comply with the NTSC or PAL standard. A typical ½-inch NTSC sensor used for a CCD CCTV camera will have 480,000 pixels, as will the 1/3- and ¼-inch sensors. Since the resolution remains constant, the larger the sensor size, the larger the pixel size can be and the greater the number of photons each pixel can collect. All other things being equal, the larger the pixel the greater will be the signal-to-noise ratio (SNR), or "sensitivity" of the sensor.

One aspect of an astro-video camera that separates it from a typical DSLR or an astronomical CCD camera is the way it outputs its images. The output from an astro-video camera is an analog video output that complies with the NTSC or the PAL standard. As we said earlier, which standard you need is dependent upon your geographic location, although today a high percentage of video equipment can automatically sense and use either standard.

Video cameras constantly refresh the image that they are displaying on a TV screen at a rate of about 60 fields a second (NTSC standard) or about 50 fields a second (PAL standard). To do this, the camera splits the image into two groups of horizontal lines called fields. A field contains information from every other horizontal line. When the camera refreshes an image, it only includes the information from one field, for example, the field containing information from the odd-numbered lines. For the next refresh cycle, the camera uses the information from the remaining field that, in this example, has the information for the even-numbered lines. This process is repeated over and over. Meantime the television set displays the fields in the order received. The result is that two refresh cycles are needed to output one frame (one complete image) from a television camera like the one used for astro-video cameras. Thus, the effective frame (complete image) refresh rate for a NTSC camera is 30 frames per second and 25 frames per second for a PAL camera. This process is known as interlacing. Either frame rate is sufficient for our brains to trick us into seeing both images as one on a TV set—in other words, a complete image. For a digital display such as a computer, the two fields are combined in the computer's memory and then displayed as a complete image. Thus, its frame rate is the same as the analog TV display.

This description assumes a lot and is rather a gross oversimplification. Just keep in mind that the TV camera is constantly refreshing the image you see 30 or 25 times per second, dependent upon which television standard is being used. Also, 30 and 25 frames a second are rounded off numbers that are used for the sake of simplicity. The actual rate for the NTSC standard is 29.97 frames per second and the PAL standard can be 23.976 frames per second. If you want to know more, an Internet search "CCTV camera analog signals interlaced" produces several excellent sites that describe the process in detail.

CCTV cameras use an electronic shutter, not a mechanical shutter, such as is used on a typical DSLR. Like most cameras, exposure times with an astro-video camera are adjustable. In the case of entry-level astro-video cameras shutter times are selectable from as short as 1/100,000 of a second to as long as 1/60 of a second (NTSC) or 1/50 of a second (PAL). However, in spite of having a shutter that cannot expose longer than 1/60 of a second (NTSC) or 1/50 of a second (PAL); some electronic "trickery" is used so that current entry- level astro-video cameras can make exposures as long as 85 seconds (NTSC) to 100 seconds (PAL).

Just what is this electronic trickery? Current astro-video cameras are what are known as integrating cameras. This means that they can integrate exposures over time, producing a final image having an exposure time equal to the sum of its parts. To do this, the camera's processer uses "stacking."

Stacking

Just what is stacking? As shown in Fig. 2.1, the frames (images) being "stacked" are aligned so that each pixel for each frame is aligned with the identical pixels for the other frames. In other words, the images are stacked on top of one another.

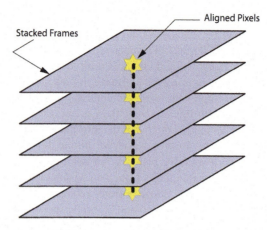

Fig. 2.1 Stacked frames

The values of each pixel stack are then either added together in what is known as a linear stack, and the total value of the stacked pixels is then used for the final image, or they are averaged in what is known as a statistical stack, and the average value of the stacked pixels is then used for the final image. (Some statistical routines use the median value of the stacked pixel.)

The photons collected by each pixel have two basic components. The first is called the signal, or simply stated, the photons that define the picture we want to take. The second is called noise. These are photons that are not part of the signal. Noise has three basic components:

- photon noise
- dark noise
- read noise.

All else equal, the signal part of an image remains constant while the noise component varies.

Photon noise is the random variation of photons reaching a camera's sensor and is particularity associated with low light imaging where signal levels are low. It creates that grainy appearance that underexposed photographs often display. Being random, photon noise has a Poisson distribution with a standard deviation equal to the square root of the mean signal value.

Dark noise is random electrons thermally generated within the camera's sensor.

Read noise is produced by the random variation of the camera's operation. Dark and read noise are also random, with a Poisson distribution. Unlike photon noise, dark and read noise are factors related to camera design.

A linear stack aligns the pixels of each image in the stack and calculates the sum of the values for each pixel stack. If the sum of the pixels in a linear stack exceeds the capacity of a pixel saturation occurs, and the excess photons will spill over into adjacent pixels, causing flaring and blooming.

The image produced by using an additive (linear) stacking process is identical to an image having the same integrated exposure time. The signal level of an image is increased directly with time, while its noise component increases as the square root of time. In other words, all else equal, a linear stack of 1024 frames having an exposure time of 1/60 of a second each produces an image with an integrated exposure time of 17 seconds that has the same signal and noise levels as will one image having a 17 second exposure time. The two images are essentially identical.

Linear stacks are typically used in integrating cameras such as the entry-level astro-video cameras this book is based upon. If blooming or flaring occurs, the number of frames being integrated is decreased. Statistical stacks are used by computer stacking programs to reduce noise in images. Because blooming or flaring cannot occur, statistical stacking allows the stacking of a large number of frames to reduce noise.

The stacking routine used by computer programs for astrophotography images to reduce noise differs from the stacking routine used by astro-video integrating cameras. For astrophotography images a statistical value is found for each stack of pixels, such as the average or median pixel value. With a statistical stack, the signal level remains constant, and noise decreases as the square root of the number of images stacked. Since the signal level does not change, the stacking process cannot saturate pixels irrespective of the number of images stacked.

The reason statistical stacking works is that the pixel values for the signal (the actual picture we want) essentially remain constant while the pixel values for noise change in a random manner. Because the signal remains constant, its average or mean value does not change. However, noise appears randomly, sometimes present and sometimes not. When it is not present, its value is zero, which drastically lowers its average or mean value. In effect, stacking averages noise out of the picture.

If a statistical stack is made using multiple copies of the same image, the values of the pixels for each image remain constant, including noise. This means that noise is treated the same as signal during the stacking process, since the noise values do not vary. If a statistical stack is done no change takes place in the signal-to-noise ratio, as the average or the median value of a set made of the same number is the number itself. For a linear stack, assuming that a linear stack does not saturate pixels, both noise and signal are increased identically, producing no change at all in the image's signal-to-noise ratio.

Compare the following three images:

- one single image having an exposure time of, say, 10 minutes with a signal to noise ratio of 3.16
- an image made from a statistical stack of 10 one-minute images having a signal to noise ratio of 1
- an image made from a linear stack of 10 one-minute images having a signal-to-noise ratio of 1.

Keep in mind that the signal part of an image is fixed while the noise component is random, following a Poisson distribution.

All other things being equal, since signal increases directly with time and noise as the square root of time, the single image having a 10-minute exposure will have a signal strength ten times that of either of the two, one-minute images and a noise level that is 3.16 times greater (3.16 is the square root of 10). Since noise is random, a statistical stack of 10, one-minute images will reduce noise by a factor of 3.16 (the square root of 10), but the signal will remain constant since the average or median value of a set containing the same numbers does not change. Although the signal-to-noise ratio of the statistically stacked image will be the same as the 10-minute image, the signal strength of the 10-minute exposure is ten times greater than the signal strength of a statistically stacked image. Even though both have the same signal-to-noise ratio, the two images are not identical.

Now take a look at the linear stack. The image resulting from a linear stack of ten, one-minute images has a signal level that is ten times greater than the signal level of a single one-minute image, while the image's noise level only increased by 3.16. The signal strength and noise of the linear stack of ten, one minute exposures is equal to the signal strength and noise of one, 10-minute exposure; thus, both images are the same.

The above discussion is a broad synopsis and simplification of the stacking process. The actual stacking process is much more involved. Just keep in mind that:

- linear stacking increases signal in proportion to time and noise as the square root of time
- statistical stacking keeps the signal constant and decreases noise as the square root of the number of images stacked
- stacking multiples of the same image will not change the signal-to-noise ratio of an image regardless of the stacking method used.

Features Found on a Typical Entry-Level Astro-Video Camera

Current entry-level astro-video cameras are typically capable of producing images having a total integrated exposure time of 85 seconds (NTSC standard) or 100 seconds (PAL standard). As discussed later, an astro-video camera has a feature called three dimensional digital noise reduction (3D-DNR). One of the noise reduction features is the capability to statistically stack up to five integrated images. A stack of five, 17-second images produces a total integrated exposure time of 85 seconds. Doing the same arithmetic for a PAL standard camera, the integrated exposure time for a stack of 1024 frames is 20 seconds, and with a total integrated exposure time of 100 seconds for a stack of five, 20-second images.

Like with most cameras, the camera operator must either select the exposure time that is used or the integration time. As with conventional cameras, shutter times are fixed. Table 2.1 provides a synopsis of shutter times and integration times typical of a budget or entry-level astro-video camera.

Modern integrating cameras have extensive noise-reduction capabilities that greatly enhance the quality of the images they produce. The camera's integrating

Table 2.1 Synopsis of entry-level astro-video camera exposure times

Shutter time (seconds)		Exposure times (seconds)	
NTSC	PAL	NTSC	PAL
1/100,000	1/100,000	0.00001	0.00001
1/10,000	1/10,000	0.0001	0.0001
1/4000	1/4000	0.00025	0.00025
1/2000	1/2000	0.0005	0.0005
1/1000	1/1000	0.001	0.001
1/500	1/500	0.002	0.002
1/250	1/250	0.004	0.004
1/120	1/120	0.008	0.008
1/60	1/50	0.017	0.02

Integrated exposures		Integrated exposure time (seconds)	
Integration	Number of images stacked	NTSC	PAL
X2	2	0.03	0.04
X4	4	0.07	0.08
X8	8	0.13	0.16
X16	16	0.27	0.32
X32	32	0.53	0.64
X64	64	1.07	1.28
X128	128	2.13	2.56
X256	256	4.27	5.12
X512	512	8.53	10.24
X1024	1024	17.07	20.48

process itself can make a significant improvement of the signal-to-noise ratio of an image. Recall in the camera's integrating process that the signal is increased directly with the number of images integrated while noise increases as the square root of the number of images integrated. Thus, all else equal, an integration of 1024 images increases the signal by 1024 times and the noise by only 32 times. The end result is the signal to noise ratio of the integrated image is 32 times greater than that of an individual image. However, more importantly, the signal level is 1024 times greater. With a NTSC standard camera having a maximum exposure time of 1/60 of a second integrating 1024 images produces a final image that has an integrated exposure time of 17 seconds. A PAL standard camera with a maximum exposure time of 1/50 of a second will produce a final image having an integrated exposure time of 20 seconds. Also keep in mind that integration is done in the camera, and is a digital process.

However, just because the camera has the ability to integrate 1024 images does not mean that the capability can be used. Remember, if the sum of the pixel values of the stacked images exceeds the capacity of the pixel to process, the pixel is saturated, and photons will spill over into adjacent pixels, causing blooming and the lost of data. Typical deep space objects often only support an integration of 64–256 images before saturating a pixel stack; thus, the resulting integration times are 1.07–4.27 seconds, respectively, for a NTSC standard camera.

Even though an integration of X1024 produces an integrated exposure time of 17–20 seconds, this is not that long of a time to acquire data from some very distant, dim, deep space object, and the image is noisy. To further resolve the noise issue, current entry-level astro-video cameras have the function known in the CCTV community as 3D-DNR. This digital process is rather complex but is capable of greatly reducing noise and enhancing the images produced. (*Note:* The Astro Video Systems DSO camera calls the 3D-DNR feature "INTMUL," integration multiple.) Regardless of what it is called, this feature is optional and is selected by setting the automatic gain control (AGC) of the camera to any value other than off.

Recall the earlier discussion on dark noise and read noise. Dark noise is light and dark spots typically resulting from an overheated sensor. The spots have no correlation with surrounding pixels. Photon noise is the random variation of photons reaching a camera's sensor and is particularity associated with low light imaging where signal levels are weak.

Entry-level astro-video cameras have a "defective pixel correction" feature named "DPC," which is nothing more than dark frame subtraction. Basically the camera makes an exposure identical to the one that is being processed *except* the shutter is kept closed. This produces the same results as the process used in astrophotography with DSLR cameras where a dust cap is placed on the telescope to block out all light. The resultant image has no signal and only the hot pixels are visible. This image is known as a dark frame since it has no image except the hot pixels. It is then subtracted from the image being processed by the camera. Since the location of the hot pixels remains constant, the subtraction removes the hot pixels, leaving only the image of the object and random photon noise.

3D-DNR (INTMUL) can be set between 0 and 5, based on the number of images the camera will stack to reduce noise. For this feature, a statistical stack is performed so noise will be reduced by the square root of the number of images stacked while the signal remains constant. If five images are stacked noise will decrease by a factor of 2.24, the square root of 5, while the signal will remain essentially constant.

Setting INTMUL to a number greater than zero also implements a process to enhance the image. The value of each pixel is compared with the values of the surrounding pixel and adjusted if the pixel value falls outside some predefined limit. Some video cameras not only compare a pixel value to that of its neighbors but also to the values in the images before and after it in a stack. This process of stacking and image enhancement provides a very powerful noise reduction tool.

For an NTSC camera the 3D-DNR process produces a final image having a total integrated exposure time up to 85 seconds for a stack of five images made using an integration of X1024 with a NTSC camera (100 seconds for a PAL standard camera). See Table 2.2. If the dark frame subtraction (DPC) option is selected, the time required for the 3D-DNR process is approximately doubled and can be several minutes.

Given the extraordinary low light sensitivity of these cameras, 85- to 100-second exposures are sufficient to image not only the brighter deep space objects such as the Messier objects but dimmer ones as well and to produce relatively noise-free images.

Table 2.2 Total integrated exposure times in seconds

Integration	Without 3D-DNR		3D-DNR without DPC	
	NTSC	PAL	NTSC	PAL
1X	0.017	0.02	0.085	0.1
X2	0.03	0.04	0.15	0.2
X4	0.07	0.08	0.35	0.4
X8	0.13	0.16	0.65	0.8
X16	0.27	0.32	1.35	1.6
X32	0.53	0.64	2.65	3.2
X64	1.07	1.28	5.35	6.4
X128	2.13	2.56	10.65	12.8
X256	4.27	5.12	21.35	25.3
X512	8.53	10.24	42.65	51.2
X1024	17.07	20.48	85.35	102.4

If a person is using a video telescope for real-time outreach, broadcasting, or personal viewing, this stacking process becomes essentially invisible once five images are stacked, as the updates stay stable as long as the telescope does not move or a cloud, an airplane, or a satellite does not pass by. When slewing to a new object for viewing, star trails caused by the moving telescope will be visible on the TV screen. Once the 'scope is at the new object, the image will slowly improve as new frames are added to the stack, and soon the star trails, etc., will not be noticeable. In fact, few objects require an exposure as long as 85 or 100 seconds; thus, the time required is often far less.

Keep in mind that the camera uses a 16-bit process for its noise reduction features. When the camera outputs an image as either a NTSC or PAL analog signal internally it uses a 10-bit digital-to-analog conversion process to produce the analog video signal. If the video signal is stored in a computer it must be digitized again. Current frame grabbers for the Window's platform are typically 8-bit devices. This has a negative impact upon processing images with a computer.

One aspect left to consider is the impact that the mount used with the video telescope has upon an integrating camera. An integrating camera simply stacks pixel by pixel. While some entry-level astro-video cameras have a stabilization mode, the magnitude of the adjustments they can make for any drift caused by vibration, poor tracking, or by field rotation remains limited. If the image moves or rotates, then light can spill over into adjacent pixels, reducing contrast and details. A good solid German equatorial GOTO mount that is precisely polar aligned provides the best platform and flexibility for imaging with an astro-video camera, especially for today's integrating cameras. However, alt-azimuth mounts can also produce excellent results, especially if integrations longer than X256 are avoided. There will be more about mounts in later chapters.

Older astro-video cameras did not have digital noise reduction capabilities and could integrate up to only 128–256 frames. The workaround for imaging was to use computer programs for stacking images in astrophotography, including making

Fig. 2.2 Astro-video camera features

Table 2.3 Astro-video camera features

Number	Item	Remarks
1	1.25-inch nose piece with cap	Removable
2	Camera body	14 grams; 43×43×64 mm
3	Tripod mount	1/4 inch/20 thread female
4	Video out	BNC female
5	RS485, hand controller port	Converted auto Iris port
6	Power in; 9–14VDC, 12VDC nominal, center pin positive	5.5-mm diameter plug with a 2.1-mm diameter center pin
7	OSD down key	
8	OSD enter key	
9	OSD left key	
10	OSD up key	
11	OSD right key	

what are known as dark frames and flat frames. This process was cumbersome and also required a significant amount of space to store the video files containing the image. Since the digital files were 8-bit, the quality was also adversely impacted.

Take a look at Fig. 2.2. The camera shown is typical of either a CCTV security camera or an astro-video camera. The typical entry-level camera is rather small in that it has no internal cooling. Its major features are listed in Table 2.3.

Field of View

An astro-video camera typically has a ½-inch or a 1/3-inch sensor. The ½-inch sensors are typically found on advanced cameras, with entry-level cameras using the less expensive 1/3-inch sensors. Actually the sensors are smaller, but the sizing convention for video sensors dates back to the vacuum tube days of television. The old standard measured the outer diameter of a camera sensor tube instead of the smaller size of the actual detector located inside the tube. In the digital world, an ½-inch CCD

Field of View

sensor measures 4.8×6.4 mm with a 8-mm diagonal that is actually approximately 1/3 of an inch; a 1/3-inch sensor has a diagonal size of 6 mm, which is approximately 1/4 of an inch, and a ¼-inch sensor has a diagonal of 4 mm, which is approximately 1/6 of an inch. Confusing? Yes. Makes sense? No, but it is the way things are.

Another aspect related to the television origin of astro-video cameras is the 1.33–1.0 aspect of the sensor dimensions. The length of an astro-video camera sensor is 1.33 times as long as its width.

An 8-inch f/10 Schmidt Cassegrain telescope (SCT) has a focal length of 2030 mm, and a 12-inch f/4.9 Newtonian typically found on a Dobson mount has a 1500-mm focal length. On the other end of telescope sizes, an 80-mm f/5 short tube refractor has a 400-mm focal length. With a 1/3 inch sensor like those typically used for an entry-level astro-video camera, the resultant image sizes are approximately 340X, 250X, and 70X for the 8-inch SCT, 12-inch for a Newtonian, and the 80-mm refractor, respectively. A ½-inch sensor will have smaller, but still rather large, image sizes, while a ¼-inch sensor will have even larger images.

However, image size is a poor way to describe the relationship between telescope focal length and sensor size, as the question, "Image size in relation to what?" is not answered. A better method is to define the area of the sky that is covered by an image, the field of view. This area can be described in arc minutes in diameter if it is circular or in arc minutes of length and width if it is rectangular. Since an arc minute is a known unit of measurement, it provides a way to accurately compare the variances in field of view caused by different camera senor sizes and telescope focal lengths. The Moon is also a good reference point, as it covers about 30 arc minutes of the night sky.

Assume that we make an image of the Moon with an 8-inch f/10 SCT using the following three cameras:

- digital SLR with an APS-C size sensor
- astro-video camera with a 1/2-inch size sensor
- astro-video camera with a 1/3-inch size sensor

A comparison of the images produced shows the relationship between the field of view—what the camera sees—and the focal length of a telescope. This relationship is graphically shown in Fig. 2.3. The figure contains an image of the Moon with three rectangles superimposed on top of it.

These three rectangles approximate the fields of view of images made with the three cameras using the same 8-inch f/10 SCT. The large rectangle outlines the image that the DSLR having an APS-C sensor produces. The slightly larger of the two small rectangles is the image produced by an astro-video camera having a one-half inch sensor, while the smallest rectangle outlines the image produced by an astro-video camera having a one-third inch sensor.

These images of the Moon clearly show that using a telescope such as the 8-inch f/10 Schmidt Cassegrain telescope used in these examples with ANY video camera, whether it has a ½-inch or a 1/3-inch sensor results in an extremely small field of view. The figure also shows the relative difference in field of view between a ½-inch and a 1/3-inch sensor is rather small.

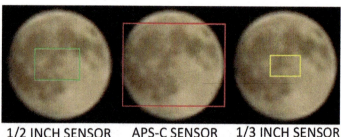

1/2 INCH SENSOR   APS-C SENSOR   1/3 INCH SENSOR

Fig. 2.3 Impact of sensor size upon field of view

Table 2.4 Astro-video camera field of view in arc minutes

Telescope focal length (mm)	¼-inch sensor (3.2×2.4 mm) Arc minutes		1/3-inch sensor (4.8×3.6 mm) Arc minutes		½-inch sensor (6.4×4.8 mm) Arc minutes		APS-C sensor (22.2×14.8 mm) Arc minutes	
	Length	Width	Length	Width	Length	Width	Length	Width
200	55	41	83	62	110	83	382	254
300	37	28	55	41	73	55	254	170
400	28	21	41	31	55	41	191	127
500	22	17	33	25	44	33	153	102
600	18	14	28	21	37	28	127	85
700	16	12	24	18	31	24	109	73
800	14	10	21	15	28	21	95	64
900	12	9	18	14	24	18	85	57
1000	11	8	17	12	22	17	76	51
1250	9	7	13	10	18	13	61	41
1500	7	6	11	8	15	11	51	34
1750	6	5	9	7	13	9	44	29
2000	6	4	8	6	11	8	38	25
2500	4	3	7	5	9	7	31	20
3000	4	3	5	4	7	6	25	17

How does all this fit into what you need to know? Two variables govern the camera's field of view, the size of a camera sensor and the focal length of a telescope. This relationship is shown in Table 2.4. A typical entry or budget level astro-video camera as well as several advanced cameras have a sensor size of 1/3 of an inch.

Note For those who are interested; the formula for calculating the field of view is:

$$fov = 57.3 * \left(\frac{d}{fl}\right)$$

where:

fov = field of view in degrees
d = sensor length or width, mm
fl = telescope focal length, mm

To convert the *fov* from degrees into minutes, as used in Table 2.3, multiply the *fov* in degrees by 60.

The diameter of the Moon is approximately 30 arc minutes. From Table 2.3 we can see that to image the Moon using a camera with an 1/3-inch sensor, the focal length of the telescope cannot be more than about 400 mm. An 8-inch f/10 SCT has a focal length of 2000 mm; thus, a 0.20x focal reducer that fully illuminates the camera's sensor is required.

Often video astronomers say that 1000 mm is the longest practical focal length for a telescope used for video astronomy. With a ½-inch sensor, a telescope having a focal length of 1000 mm has a field of view approximately half the area of the Moon. A field of view equal to about half the area of the Moon is considered by many as about the smallest field of view that is practical for capturing most deep space objects. Since the field of view of a 1/3-inch sensor is smaller than a ½-inch sensor, using the same logic implies that a telescope's focal length for a 1/3-inch sensor should not exceed around 800 mm (Table 2.4).

Since most larger telescopes used by amateurs have focal lengths longer than 1000 mm, a focal reducer must be used or the views and images produced will be akin to viewing through a soda straw. Here we enter into a strange world of stacking focal reducers. Often a 0.63x and a 0.5x focal reducer are used with SCTs to obtain a final reduction of around 0.3x. Fortunately the small sensor size of a 1/3-inch sensor helps in this quest, as the very small sensor area typically remains illuminated. Variable reducers are also popular and are available from some astro-video camera makers. One variable focal reducer with the included spacers and extension tubes achieves nominal ratios of 0.63x, 0.56x, 0.5x, 0.43x, 0.36x, and 0.29x. One camera maker even has a 0.18x focal reducer for Meade and Celestron f/10 SCTs. Also in demand but difficult to find is the 0.33x SCT focal reducer, which was produced by Meade at one time.

Consider M42, the Great Nebula in Orion. It has an apparent size of 60×90 arc minutes. To capture the entire nebulae, you will need a telescope having a focal length of around 200 mm with either a 1/3-inch or ½-inch sensor. Recall that available focal reducers for 1/3-inch camera sensors go down to about 0.30x. This means that the longest focal length telescope that can be used to capture the Great Orion Nebula in its entirety with a 1/3-inch astro-video camera is about 700 mm if a 0.3x focal reducer is used. On the other hand, M1, the Crab Nebula, has an apparent size of 6×4 arc minutes. Just about any telescope including Schmidt Cassegrain telescopes with apertures up to 12 inches can capture the entire nebula with no difficulty.

There is no law that says you have to have the entire object in the field of view of your camera. The above discussion is to make you aware of how your camera will see the universe and how you can change what it sees. However, there are some practical considerations. A small field of view increases demands upon the stability and accuracy of a telescope's GOTO mount in the same way it does for an optical telescope.

Image Brightness

Recall the definition of the focal ratio of a telescope—the focal length divided by the telescope's aperture. A focal reducer works by optically shortening the focal length of a telescope, thus lowering the focal ratio of the telescope at the same time. This has another impact to consider—image brightness. If we use a 0.63x focal reducer on an f/10 telescope, we will also reduce the focal ratio from f/10 to f/6.3. This means the image will be brighter; thus, we will be able to image fainter parts of an object, fainter objects, etc., or we can shorten our integration time. (For photographers making the transition to astronomy, please note that in astronomy the term aperture refers to the diameter of the primary mirror or lens of a telescope and is fixed.)

Let's take a look at two telescopes, an 8-inch f/10 SCT with a 0.5x focal reducer and an 80-mm f/5 refractor with no focal reducer. The focal ratios of both telescopes are the same (f/5.0), so the image brightness is the same. However, the focal length of the 8-inch SCT is now 1000 m and its image with a 1/3-inch sensor has a field of view of 17×12 arc minutes, while the focal length of the refractor is 400 mm and its image with a 1/3-inch sensor has a field of view of 41×31 arc minutes. Even though the images in both telescopes are equally bright, the image in the 8-inch SCT is considerably larger, with more visible detail than that produced by the 80-mm refractor. Now if we shortened the focal length of the 8-inch f/10 telescope to 400 mm, the same as the 80-mm f/5 refractor, the focal ratio of the SCT becomes f/2. Although the image sizes and field of view will be the same, the image in the f/2 SCT is far brighter than that in the f/5 refractor.

Consider that for most distant deep space objects the number of photons per second an object sends our way essentially remains statistically fixed over the relatively short period of time that a typical observing session lasts. The total number of these photons that enter into a telescope over any given period of time is constant per unit area of a telescope's light collection area (its primary mirror or lens). Thus, the larger the aperture of a telescope, the greater the total number of photons it can collect. This would seem to indicate that the larger the aperture of a telescope, the better it is for photography. However, as we shall soon see, this is not altogether true.

Recall, one parameter that describes a telescope's optical train is its focal ratio—a measure of image brightness. This has a direct impact upon photography, as the brighter the image the more photons that are concentrated on the camera's sensor over any given time—thus, the higher the signal to noise ratio or the shorter the resultant exposure time. In other words, the lower the focal ratio of a telescope, the higher its exposure efficiency for photography.

In many cases exposure efficiency of a telescope is not that important because individual exposure times can be changed to match the telescope. All that happens is that the total amount of time needed to obtain an image changes. However, with lightweight azimuth and equatorial mounts a limit exists for the length of exposure times due to field rotation, mount stability, and, for equatorial mounts, the preciseness of the polar alignment. Thirty seconds with no movement of the telescope due to field rotation, tracking, or vibration is about as good as it gets with an alt-azimuth mount. Integration times longer than this risk softer images and larger stars. This implies that an astro-video camera, with digital noise reduction enabled, is limited to an integration of around X256 with digital noise reduction enabled and X1024 without digital noise reduction. These integration times are sufficient for a wide range of deep space objects. This also implies that for video astronomy, the mount and tripod are not as important as they are for traditional astrophotography.

Integrations even as high as X1024 do not have time to collect many photons, so the signal to noise ratio will not be that great. Fortunately, for bright deep space objects, like most of the Messier and Caldwell objects, sufficient photons can be captured by a telescope on a lightweight azimuth or equatorial mount to produce nice images. As objects become dimmer, the number of photons captured by the telescope and camera decreases, and the signal-to-noise ratio of each frame decreases. A point is reached where the signal strength is too weak for noise reduction methods to address.

Consider two telescopes, one with a focal ratio of f/5 and the other with a focal ratio of f/10. The telescope with a focal ratio of f/5 will have a brighter image than the telescope with a focal ratio of f/10. Over any given period of time the f/5 telescope will concentrate 4 times as many photons per unit area of the camera's sensor than a telescope with a focal ratio of f/10. Adding a 0.63x focal reducer to a f/10 telescope improves its exposure efficiency. However, the f/5 telescope will still be almost twice as efficient, collecting 1.6 times as many photons per unit area of the camera's sensor than the f/10 telescope with a 0.63x focal reducer.

Refractor and Newtonian telescopes in use today can have focal ratios as low as f/4, while nearly all Schmidt Cassegrain telescopes sold today have a focal ratio of f/10, and the focal ratios of Maksutov Cassegrain telescopes typically range from f/12 to f/15. The high focal ratios of Maksutov and Schmidt Cassegrain telescopes produce dimmer images that require longer exposures than telescopes having a low focal ratio. Because telescopes with high focal ratios need longer exposures they are often referred to as slow telescopes, and telescopes with low focal ratios are referred to as fast telescopes.

Telescopes with high focal ratios typically have long focal lengths. For video astronomy cameras with their small sensors, this means they have large but dim images. The performance of both Schmidt Cassegrain telescopes and Maksutov Cassegrain telescopes can be enhanced by attaching a focal reducer to the telescope. One typical focal reducer available today for Schmidt and Maksutov Cassegrain telescopes theoretically reduces their effective focal length to 63% of the original length. In reality, the actual result may differ a little. Regardless, the resultant impact upon exposure times is significant, as shown by Table 2.5.

Table 2.5 The impact of focal ratio on equivalent exposure times

Focal ratio	Exposure time (seconds)	Remarks
f/2.0	0.25	
f/3.0	0.5	
f/4.0	1.0	
f/5.0	1.6	
f/6.3	2.5	f/10 telescope with a 0.63 focal reducer
f/7.6	3.6	f/12 telescope with a 0.63 focal reducer
f/8.1	4.0	f/13 telescope with a 0.63 focal reducer
f/8.8	4.9	f/14 telescope with a 0.63 focal reducer
f/9.5	5.7	f/15 telescope with a 0.63 focal reducer
f/10	6.3	
f/12	9.0	
f/13	10.6	
f/14	12.3	
f/15	14.0	

Although these exposure times are related to astronomical cameras they will impact the integration times selected for an astro-video camera in a very similar manner. For example, the total integration time required for a telescope at f/8.1 is 4 times the integration time required for a telescope at f/4.0.

For the mathematically inclined, here is the equation for calculating exposure times.

$$E_n = E_o \left(f_n / f_o \right)^2$$

where:

E_n is the new exposure time
E_o is the old exposure time
f_n is the new focal ratio
f_o is the old focal ratio.

Astro-Video Camera Settings and Adjustments

Two categories of cameras are typically thought of as budget astro-video cameras:

- stock CCTV security cameras used by experimenters who modify them
- commercially available astro-video cameras derived from stock CCTV cameras used by people interested only in observing, broadcasting, or imaging.

The same CCD sensor developed by Sony Corporation dominates both categories, specifically the Sony EX VIEW HAD II CCD sensor. The EX VIEW HAD II

Astro-Video Camera Settings and Adjustments

Fig 2.4 OSD menu bar

Table 2.6 OSD menu titles

CCTV security camera	Astro-video camera A	Astro-video camera B
EXPOSURE	AVSYSTEMS	EXPOSURE
COLOR	COLOR	COLOR
DAY NIGHT	DAY AND NIGHT	DAY NIGHT
FUNCTION	EFFECT	EFFECT
MOTION DETECTION	MOTION	MOTION
MASKED AREA	TEST BARS	PRIVACY
SET	PROCAMP	ENHANCE
VIDEO CAMERA	SYSTEM	SYSTEM
EXIT	EXIT	EXIT

CCD sensor is available in three sizes, ½-inch, 1/3-inch, and ¼-inch. Since essentially all the cameras are based upon the same CCD camera sensor, the basic camera settings and controls are for the most part independent of the camera brand. However, CCTV and commercial astro-video camera companies do have some minor differences in their specifications, so their settings do vary somewhat. One of the major differences is simply assigning a different name to the same feature.

Current entry-level astro-video cameras are controlled using a pull-down menu that is displayed on the television monitor or, if one is used, a computer. This display is called an OSD (on screen display). The menu system used for the OSD as well as how the menus are accessed and used is essentially identical for all the current entry-level cameras (Fig. 2.4). The same is true for most, not all, of the features and settings available, as camera manufacturers do use their own programming. In any case, the differences apparent to the user are not that numerous. Once someone knows and understands the controls and settings for one camera brand, setting up and using cameras from other manufacturers is doable with little to no difficulty. See Table 2.6.

To simplify the discussion of camera settings, the DSO-1 camera produced by Astro Video Systems will be used as an example only because it is the camera this author is most familiar with. Other cameras will have similar, almost identical, settings.

Access to the camera's settings and controls is simple. Look at the back of the camera (Fig. 2.5). There you will see five buttons. Some cameras have a remote

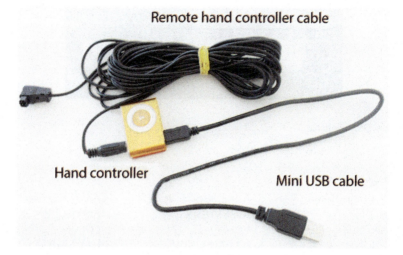

Fig. 2.5 Camera remote hand control

Fig. 2.6 AVSYSTEM menu

hand control with the five buttons. Press the center button. A horizontal bar will appear across your screen. This is the OSD menu bar. The OSD menu bar has the following nine categories:

- AVSYSTEMS (see Fig. 2.6, AVSYSTEM menu)
- COLOR (see Fig. 2.7, Color menu)
- DAY AND NIGHT (see Fig. 2.8, Day and Night menu)
- EFFECT (see Fig. 2.9, Effect menu)
- MOTION (see Fig. 2.10, Motion menu)
- TEST BARS (see Fig. 2.11, Test Bars)
- PROCAMP (see Fig. 2.12, Procamp)
- SYSTEM (see Fig. 2.13, System)
- EXIT (see Fig. 2.14, Exit)

Astro-Video Camera Settings and Adjustments

Fig. 2.7 Color menu

Fig. 2.8 Day and Night menus

Fig. 2.9 Effect menu

Fig. 2.10 Motion menu

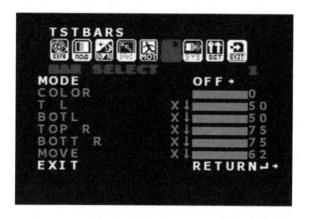

Fig. 2.11 Test Bars

Fig. 2.12 Procamp menu

Astro-Video Camera Settings and Adjustments

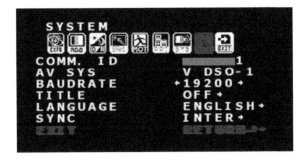

Fig. 2.13 System menu

Fig. 2.14 Exit menu

Navigation through the OSD menu bar is done using the four buttons surrounding the center button located on the back of the camera or the hand controller. The center button is similar to the Return key on a computer keyboard. A few minutes practicing with the menu bar are generally more than sufficient to gain proficiency.

AVSYSTEM Menu Options

The AVSYSTEMS menu contains most of the commands and settings of interest for astronomy. It has six major categories:

- Exposure
- Sense up
- AGC
- INTC
- Brightness
- Exit

Exposure has four modes: Normal, BLC (backlight compensation), HLI (highlight), and WDR (wide dynamic range). The normal mode is typically used for most viewing applications and is a good default exposure mode to select. It has no options for adjustments. While back light compensation has little application for astronomy, it could be useful for terrestrial work. The amount of compensation is adjustable via a slider control with values ranging from 0 to 3.

The highlight mode is adjustable via a slider control with values ranging from 0 to 40. This mode is useful for solar viewing to mask the sun disk.

The wide dynamic range mode has two options: "auto" and "on." If the "auto" option is selected, the range is controlled by a slider control having values ranging from 1 to 20. If the "on" option is selected, the camera will automatically adjust the dynamic range of the camera. This setting is useful for objects such as the Moon.

Sense up is one way to set shutter speed. Sense up is set to "OFF" and not used for astronomy, as it consumes sufficient electrical power to create amplifier glow in images. Available integration and shutter speeds are: X1024, X512, X256, X128, X64, X32, X16, X8, X4, X2, AUTO, 1/60, 1/100, 1/120FLC, 1/250, 1/500, 1/1000, 1/2000, 1/4000, 1/10,000, 1/100,000. The 1/120FLC setting controls flicker.

Note These shutter speeds are for a camera built to the NTSC standard. The shutter speeds for a camera built to the PAL standard are identical except that the exposure time immediately after "AUTO" will be 1/50th of a second instead of 1/60 of a second.

Automatic gain control (AGC) has a dual role. It is similar to the ISO settings used by DSLR cameras. The signal is amplified electronically. Four settings are available: off, 12 dB, 24 dB, and 36 dB (off, low, medium, high). As with a DSLR camera, noise is dependent upon the amount of gain selected, with 36 dB producing the most noise. For general viewing, AGC is set to "off." If the camera's 3D digital noise reduction features are desired for usage, then the AGC must be set to any value other than OFF, preferably 36 dB (high).

The INTC setting is where electronic shutter speeds or integrated exposure times are selected. Two options are provided, auto and manual. Select manual. The available exposure times or shutter speeds for manual selection are: X1024, X512, X256, X128, X64, X32, X16, X8, X4, X2, AUTO, 1/60, 1/100, 1/120FLC, 1/250, 1/500, 1/1000, 1/2000, 1/4000, 1/10,000, 1/100,000. The automatic option (DS/IRIS) will show on the display, but the IRIS port of astro-video cameras is typically disabled and used for the remote hand controller connection.

Here are some things to note:

- These shutter speeds are for a camera built to the NTSC standard. A camera built to the PAL standard will have the same shutter speeds except the shutter speed of 1/50 of a second is used instead of 1/60 of a second.
- The maximum exposure time is 17 seconds for NTSC. The times for a PAL sensor are very similar and have a maximum exposure time of 20 seconds.
- For some cameras, the automatic option (DS/IRIS) is disabled.

Astro-Video Camera Settings and Adjustments   31

Brightness allows you to set the brightness of the image you see on your screen. Brightness is adjustable using a slider control having values ranging from 0 to 99.

You can toggle between return, exit, or save/exit. Return will return you to the previous menu. Exit will close the menu without saving any changes. Save/exit closes the menu but saves any changes made.

Color Menu Options

This menu is for adjusting the color balance of the images made by the camera. It has five options:

- AWB (auto white balance)
- ATW (auto tracking white balance)
- Manual
- Push
- Exit/Return/Save&Exit

AWB automatically determines white in a captured image and then adjusts the white balance accordingly. Two slider controls with a range of 0–255 are available to manually adjust the red gain and blue gain.

ATW is almost identical to AWB except instead of using a captured image, it continuously monitors and adjusts for white balance. As with the AWB option, two slider controls with a range of 0–255 are available to manually adjust the red and blue gains.

Manual allows setting the white balance by adjusting red and blue saturation levels. Two slider controls, one for red and the other for blue, are available for manual saturation adjustments within a range from 0 to 255.

The Push setting is an automatic color balance option. When Push is selected, the camera will automatically measure and adjust the color balance. No options exist for manually adjusting this setting.

You can toggle between Return, Exit, or Save/Exit. Return will return you to the previous menu. Exit will close the menu without saving any changes. Save/Exit closes the menu but saves any changes made.

Day and Night Menu Settings

The Day and Night menu controls whether the camera output is a color or monochrome signal. Five options exist as follows:

- Day
- Night
- Auto
- Smart LED
- Return/Exit/Save&Exit

If the Day option is selected, the camera's output is in color. Color settings are adjusted in the Color menu.

The Night option sets the camera to a monochrome output with an option called Burst to enhance the image. The Burst option can cause flaring on bright objects and also can prevent the camera from switching to the monochrome mode.

With the Auto option the camera evaluates the image based upon its intensity. Two slider controls are available to set the changeover points from Day to Night or from Night to Day. A delay time setting is also provided as well as the Burst option. The Auto option is not applicable for astronomical uses.

The Smart LED option is very similar to the Auto option except a photo cell is used to determine when to switch from Day to Night or Night to Day. This option is often disabled. The Smart LED option is not applicable for astronomical uses.

You can toggle between Return, Exit, or Save/Exit. Return will return you to the previous menu. Exit will close the menu without saving any changes. Save/Exit closes the menu but saves any changes made.

The Day and Night options are of interest for astronomical purposes. It is important to think in terms of do I want a color image or do I want a monochrome image whether or not you are imaging during the day or night. Simply substitute in your mind the word "Color" for "Day" and "Monochrome" for "Night," as you can certainly use either the color mode or the monochrome mode at night or during the day. The Day/Night terminology is derived from the security industry and has no relevance to astronomical usage.

Effect Menu Settings

The Effect menu governs how an image is presented on either a television screen or a computer monitor. It has the following options:

- Freeze
- Mirror
- Digital Zoom
- Sharpness
- Stabilizer
- Return/Exit/Save&Exit

The Freeze option has two states, either "on" or "off." The option does exactly what is says—it freezes the display on the screen when set to "on."

The Mirror option has four sub options: "off," Horizontal," "Vertical," or "Rotate." These sub options do exactly what they say; Horizontal switches the screen in the horizontal orientation, Vertical in the vertical orientation, and Rotate rotates the image by 180 degrees. This option is useful to adjust for the reverse images of SCTs and inverted images of reflectors.

The Digital Zoom option will optically enlarge the image. The enlargement is set by a slider control that provides for an optical zoom ranging from 1 to 10x. Many people find that using the optical zoom is an excellent aid for bringing the camera to focus.

The Sharpness option can help improve the definition of an image. The degree of sharpness is governed by a slider control having values from 0 to 50. A value of 0 is essentially the image as captured by the camera. Increasing the value increases the sharpness of the image. It also increases any artifacts in the image, and can cause halos around stars. A good starting value is between 15 and 25.

The Stabilizer is the last option available. It is designed to reduce screen judder and is either set to "on" or "off." When set to "on" the Digital Zoom option is disabled. The stabilizer option is very helpful viewing or imaging planetary objects. It also helps keep stars from dancing around on nights with poor seeing conditions. *Note:* the stabilizer feature and the zoom feature use common resources in the camera; if one is selected, the other is disabled.

You can toggle between Return, Exit, or Save/Exit. Return will return you to the previous menu. Exit will close the menu without saving any changes. Save/Exit closes the menu but saves any changes made.

Motion Menu Settings

The Motion menu has no astronomical applications at this time. It has the following options:

- Motion Detection
- Alarm Mode
- Sensitivity
- Area Select
- Output Time
- Auto Zoom
- Return/Exit/Save&Exit

The Motion Detection option is set to either "on" or "off." The Alarm Mode is set to either "off," "Message," or "Area." The Sensitivity option has a slider control ranging from 0 to 120 to adjust for movement sensitivity. Up to four areas can be designated by the Area Select option. The Output Time controls the alarm time with a slider control adjustable from 0 to 10. Auto Zoom is either set to "on" or "off."

You can toggle between Return, Exit, or Save/Exit. Return will return you to the previous menu. Exit will close the menu without saving any changes. Save/Exit closes the menu but saves any changes made.

Test Bars Menu Settings

The Test Bar menu is meant to provide privacy by masking out areas of the camera's view such as a bedroom window in an apartment across the street from the camera. For astronomy purposes, the mask can be configured to form a cross that can be used to center stars either while aligning a telescope mount or doing a precise polar alignment.

Procamp Menu Settings

The Procamp menu has the following six settings:

- Gamma
- INTMUL (3D-DNR)
- Monitor
- DPC
- Vignetting Correction
- Return/Exit/Save&Exit

Gamma controls the image brightness. The control has four settings; 1.0, 0.6, 0.45, and 0.3. A setting of 1.0 produces the darkest image while 0.3 produces the brightest image. No one setting is better than the other. Select the one that is most pleasing.

INTMUL is a noise reduction feature of the camera typically known as 3 Dimensional Digital Noise Reduction (3D-DNR) in the CCTV community. For this feature to work, the AGC option in the AVSYSTEMS menu must be set to any value other than "off." A setting of 36 dB is recommended. The INTMUL value should be set to 5 for the greatest noise reduction.

For the best results always select five images. The camera does not align the image based upon stars or other features. The alignment is strictly an alignment by pixels. There can be no drift or movement of the camera. If the total integrated time exceeds about 30 seconds, then an equatorial mount will likely be needed to counter field rotation.

The Monitor option adjusts the display for either a CRT monitor or an LCD screen. Select CRT if you have an older computer monitor or television set with a cathode ray tube. Select LCD if you have an LCD monitor or television.

DPC is set to either "Auto" or "off." When set to Auto, the camera will attempt to eliminate hot pixels using dark frame subtraction.

Vignetting Control has a slider control with a range of between 0 and 50. It is used to adjust peripheral brightness of an image.

You can toggle between Return, Exit, or Save/Exit. Return will return you to the previous menu. Exit will close the menu without saving any changes. Save/Exit closes the menu but saves any changes made.

SYSTEM MENU settings

The System menu sets parameters required for the camera to communicate with a computer. The menu has the following seven items:

- Comm ID
- AV System
- Baudrate
- Title
- Language

- Sync
- Return/Exit/Save&Exit

The communications ID is a number assigned to each camera in a security network. It is set by a slider control having values ranging from 0 to 254. This feature is only useful if more than one camera is connected to a computer.

AV System is permanently set to V DSO-1. This is the communications protocol that determines how the computer and camera will communicate with one another. It is fixed and no adjustment is possible.

Baudrate has a slider control to set the baudrate used for communications between the camera and computer. Six rates are available: 2400, 4800, 9600, 14,400, 19,200, and 38,400. A setting of 19,200 is a good default setting.

Title is a feature not typically used in astronomy. It is set to "off." It can be used to provide a "text" identification on the image. To use it select "on." An edit box will appear similar to a keypad on a smartphone where you can enter text and numbers.

Language offers two options—English or Chinese. Do not set to Chinese unless you are proficient in that language!

Sync sets internal synchronization to either internal or automatic. Always set to internal.

You can toggle between Return, Exit, or Save/Exit. Return will return you to the previous menu. Exit will close the menu without saving any changes. Save/Exit closes the menu but saves any changes made.

Exit Menu Settings

The Exit menu has the following four options:

- Exit
- Exit&Save
- Initialize
- Return

Exit will close the OSD menu screen without saving any changes that were made. Save/Exit saves all changes and then closes the OSD menu screen. Initialize resets the camera to default settings. WARNING: "Initialize" also resets the language to Chinese. Unless you are proficient in the Chinese language, the "Initialize" setting should not be used. Return returns to the previous menu.

How Does a Video Camera Work?

Integrating Cameras

Video cameras are not really that complex to use. Once you understand how they work the settings become readily apparent. Here is a synopsis of some major attributes of typical entry-level astro-video cameras to keep in mind.

The first thing one must do is forget just about everything you know about how cameras work. You are using a camera that operates on a completely different principle than a DSLR or an astronomical CCD camera. A typical camera has an exposure time set by the operator. When the shutter is tripped, the camera will make one image with an exposure time equal to that set by the operator. You set the camera at 20 seconds and you will get one exposure that is 20 seconds long.

With an astro-video camera, the NTSC standard limits the exposure time to 1/60 of a second (0.0167 seconds). While an exposure time shorter than 0.0167 seconds is possible, no exposures can be longer. Now let us look at the process if we want an exposure time that is greater than 1/60 of a second with an astro-video camera. For an example, if we want a 17-second exposure we cannot open the shutter for 17 seconds, as the slowest speed it can capture is 1/60 of a second (0.01660 seconds). This may lead to the conclusion that a 17-second exposure is not possible with an astro-video camera. However, this is not the case. Recall that modern astro-video cameras are known as integrating cameras. The integrating process is nothing more than a linear stacking of the 1/60 of a second exposures to produce a final image equivalent to the sum of the exposure times of the images stacked.

Look at the problem above. We want a 17-second exposure with a camera that is limited to a maximum exposure time of 0.0167 seconds. This is where the integration capabilities of an astro-video camera come into play. If we divide 17 seconds by 0.0167 seconds per image, we find that total exposure time of 17 seconds will contain 1024. In other words, integrating (stacking) 1024 images having an exposure time of 1/60 of a second each gives an integrated exposure time of 17 seconds. This is written as X1024. Current budget-level integrating cameras can do the following integrations:

X2 (0.03 seconds)
X4 (0.07 seconds)
X8 (0.13 seconds)
X16 (0.27 seconds)
X32 (0.53 seconds)
X64 (1.07 seconds)
X128 (2.13 seconds)
X256 (4.27 seconds)
X512 (8.53 seconds)
X1024 (17.07 seconds)

In addition to producing the integrated images, the 3D-DNR feature of current astro-video cameras can stack up to five of the integrated images typically using a statistical stacking process. For an NTSC camera this means that the final image produced by the camera can have a total integrated exposure time of up to 85 seconds (5 times (X1024) or 100 seconds for a camera made to the PAL standard.

Astro-Video Camera Screen Refresh

Astro-video cameras put out an analog television signal that complies with the NTSC or the PAL standard. The video cameras constantly refresh the image that they are displaying on a TV screen. The camera divides the image into two fields. For the NTSC standard one field is displayed every 1/60 of a second followed immediately by the second field. The television set displays the frames in the order received. The result is that two refresh cycles are needed to output one frame (a complete image) from an astro-video camera. This means the effective refresh rate for an NTSC camera is 30 frames per second (25 frames per second for a PAL camera). This process is known as interlacing.

Because of the display rate, our brains trick us into seeing both images at the same time on a TV set—in other words, a complete image. For a digital display such as a computer, the two fields are combined in the computer's memory and then displayed. This description assumes a lot and is a gross oversimplification. Just keep in mind that the TV camera is constantly refreshing its image at a rate of 25 or 30 times per second dependent upon which television standard is being used.

The next factor to consider is the impact that the camera's frame rate has on an integrating camera. With an NTSC camera, an integration of say X128 takes 2.13 seconds. During this 2.13-second integration period the camera will display 64 identical images before displaying the integrated image. The process will then repeat for the next integration cycle and so on. This has some implications if the astro-video camera is used to image deep space objects, such as those that are discussed in Chap. 7.

Again, this synopsis makes many assumptions and is intended as a broad overview only. If you want to know more, an Internet search "cctv camera analog signals interlaced" produces several excellent sites that describe the process in detail.

General Notes Regarding Entry-Level Astro-Video Cameras

- 3D-DNR (INTMUL) should be set to 5 and DPC set to off unless you have an excessive number of hot pixels. Automatic gain control (AGC) should be set to 36 dB, Gamma set to 1, and Brightness to around 60 but always higher than 30. If an image is too bright, reduce brightness but not to less than 30. If that does not work, then reduce the integration; do not reduce the automatic gain control.
- 3D-DNR should never be disabled. Although not expensive, the current entry-level astro-video cameras have extensive sophisticated noise reduction capabilities, and the camera's processor uses 16 bits for processing. The camera's analog-to-digital conversion uses 10 true bits per pixel (10 bits for cyan, 10 for green, 10 for magenta, and 10 for yellow). Currently, frame grabbers for the Windows Platform are 8-bit devices.

- With 3D-DNR (INTMUL) set to 5, an integration of say X256 produces an image equivalent to an exposure of 20 seconds (5×256/60=21). The maximum integration of X1024 is 87 seconds (NTSC) and 100 seconds (PAL).
- If you are using an alt-azimuth mount, choose an integration of X256 or less with 3D-DNR enabled. This produces a 20-second integration, which, for most of the sky, is short enough so that field rotation is not obvious in an image. If an integration of 256 is not sufficient, then your image will not be as sharp as one made with an equatorial mount.

A Sample of Currently Available Entry-Level Astro-Video Cameras

Here is a listing of four companies that offer commercially available cameras suitable for astro-video astronomy that cost around $100. All these cameras are based upon CCTV technology and have analog outputs. Two of the companies have done extensive reworking, including the complete replacement of some circuit boards and firmware programming. The cameras are listed in order of the most modified down to the least modified. All four of these companies responded to a request for information.

Bear in mind that some of these companies are very small and have limited production capabilities. Delivery times vary. There are other companies that have websites that make similar products. However, they did not respond to a request for information and are not included on the list. Also, cameras are available from the major telescope manufacturers and astrophotography camera makers. These cameras are excellent but were digital cameras, fell outside the price range, etc., and are not listed.

Astro Video Systems

http://www.astro-video.com/
Camera: DSO-1
Sensor: Sony 1/3-inch 960H EXview HAD CCDII

- Camera only: $110
- DSO entry-level system: $180
- DSO-1 camera
- OSD hand remote
- BNC to RCA adapter
- 1.25-inch nosepiece
- 20-foot video and power cable
- 20-foot remote control cable
- 2-m USB cable
- Battery adapter

MallinCam

http://www.mallincam.net/
Camera: Micro EX
Sensor: Sony 1/3-inch 960H EXview HAD CCD II

- Camera only: $100
- Micro-EX System: $170
- Micro-EX camera body
- C-mount to 1.25-inch eyepiece adapter
- AC power supply
- 25-foot video/power cable
- BNC to RCA adapter

Revolution Imager

http://www.revolutionimager.com/
Sensor: Not specified
Revolution Imager System R2: $300

- CCTV camera
- Portable 7" LCD monitor
- 12v Li-ion battery with charger to run both the camera and LCD screen
- 0.5x focal reducer
- UV/IR filter
- Hand-held camera remote control
- Shock-proof carry case
- All required cables

Lntech

http://www.lntech.com.cn/
Camera: LN300
Sensor: Sony 1/3-inch 960H EXview HAD CCD II
Price: camera only: $71
http://www.aliexpress.com/snapshot/284874236.html (unmodified CCTV security camera)

Chapter 3

Assembling Your Video Astronomy Kit

Getting Started

Let's take a look at a budget-priced video telescope system designed to visually observe the night sky. Recall earlier that we said all that was needed to convert an optical telescope into a video telescope was to simply replace the eyepiece with a TV camera connected to a TV set. This is true, but like everything in life there are some details to consider, and some components perform better than others.

The basic video telescope is set up to visually view the night sky only. Like viewing through an eyepiece, no images are processed or stored for future viewing. What you see on your TV monitor screen is what you get. For most deep space objects the view is essentially real time, with a typical delay of 10–20 seconds at the most. However, for some extremely dim objects the delay can be around 90 seconds. The basic video telescope that is only used to visually observe the night sky is rather simple. Here is a list of the essential components:

- Telescope with finder
- Telescope mount and tripod (tracking or GOTO)
- Astro-video camera
- TV monitor, TV set, or DVR player
- Ancillary components
- Focal reducer
- Eyepiece(s)
- Power supply (supplies)
- Power cable(s)
- Camera OSD hand controller

- Camera OSD hand controller cable
- Video cable
- BNC to RCA adapter

If imaging, broadcasting, saving the viewing session, etc., are objectives, then the following additional items are required:

- Video capture device (frame grabber)
- Labtop computer or Win10 tablet
- Video software
- External memory (optional)

Astro-Video Cameras

Probably the first choice to make is deciding which camera you want to use. For someone who is just beginning or who does not want to risk a sizable amount of money, three entry-level/budget astro-video cameras are currently available in complete kits that have prices that are very competitive with quality eyepieces:

- Astro Video Systems DSO-1 camera
- MallinCam Micro EX camera
- Revolution Imager Revolution camera

These cameras are available as a standalones for around $100 US or around $200–300 as a bundled kit containing the camera and the accessories necessary for an amateur to set up a video telescope for viewing the night sky. Each manufacturer also has LCD TV monitors available as well as other accessories. These three entry-level cameras are based upon the same CCD technology and are very similar in features and performance. Their low price and capabilities are enticing many experienced astronomers to join the rapidly growing ranks of video astronomers.

Don't let the low price of these three entry-level astro-video cameras fool you. They are very capable cameras. The major differences between them and more expensive cameras are:

- Noise (cooling)
- the size of the camera sensor
- integration times.

Internal cooling is one feature the manufacturers had to cut to produce a camera for around $100. The lack of internal cooling does impact these entry-level cameras, especially in warm climates. However, although their images are notably noisier in hot weather than in cold weather, they are very usable. Their small sensor size creates minor issues related to finding and fitting objects in the camera's field of view. While their integration times are less than higher priced cameras, all three can do an X1024 integration. This is more than adequate to capture most objects in the night sky. All in all, the performance of these entry-level cameras is very competitive with the more pricy cameras currently on the market.

Astro-Video Cameras 43

Table 3.1 Telescope focal length and image size

Telescope focal length (mm)	1/3-inch sensor image size (4.8×3.6 mm)			1/2-inch sensor image size (6.4×4.8 mm)		
	Field of view Arc minutes			Field of view Arc minutes		
	Length	Width	Magnification	Length	Width	Magnification
200	83	62	33	110	83	25
400	41	31	67	55	41	50
600	28	20	100	37	28	75
800	20	16	133	28	20	100
1000	17	13	167	22	17	125
1250	13	10	208	17	13	156
1500	11	8	250	14	11	188
2000	8	6	333	11	8	250
3000	5	4	500	7	5	375

From Chap. 2 recall the small sensor chip sizes used with astro-video cameras. The diagonal of a typical entry-level astro-video camera is 6 mm, while that used in more advanced cameras have a diagonal of 8 mm. These small sensors can produce some rather large image sizes with a narrow field of view, as shown in Table 3.1. This is also demonstrated by Fig. 3.1.

A digital output is not required for visual work. The analog video signal output from an astro-video camera can feed any TV set, TV monitor, DVD player, DVD recorder, VCR, camcorder, etc., having an analog video input. Fortunately most still do. Some amateurs simply record their night's session on a DVD recorder, VCR, etc., and play back for viewing and study later. Although HDTV converters can make the camera's output compatible with HDTV devices, the image resolution will remain the same as before being converted.

Not all applications of video astronomy use an analog video signal. If the camera will be used to broadcast on the Internet, save and process images on a computer, etc., a digital signal is required. This requires a video capture device commonly called a frame grabber.

Earlier, mention was made of accessories needed to connect an astro-video camera to a telescope for viewing the night sky with a TV, LCD video monitor, DVD player, Camcorder, and so on. As shown in Fig. 3.2, the list actually is quite short:

- Hand controller cable
- Camera
- Video coax RG 59/power cable
- C mount with 1.25-inch nosepiece
- C mount dust cap
- USB cable
- USB video grabber
- BNC female to RCA male adapter
- Camera remote hand controller
- Camera dust cap

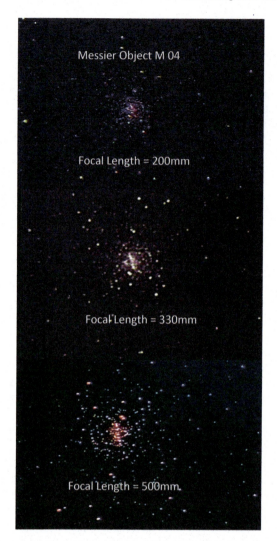

Fig. 3.1 M04 at 500-, 330-, and 200-mm focal lengths

If broadcasting, image enhancement, data storage, or any other activity that needs a computer is on your list of requirements, then the analog output signal of these three entry-level cameras must be converted to a digital signal. To do this, a video capture device (a frame grabber) is needed along with a computer or tablet with the appropriate software and memory. All of the camera manufacturers sell frame grabbers for their cameras at prices competitive with generic brands found on the Internet. Just about any computer will do, including Windows 10 tablets.

Fig. 3.2 Astro-video camera accessories

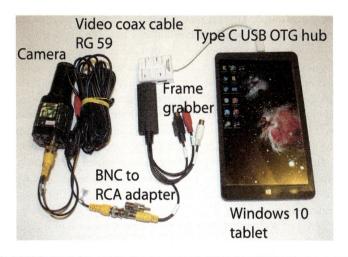

Fig. 3.3 Analog-to-digital USB conversion

The computer must have the appropriate software and sufficient memory. A tablet must also have an USB port such as the Type C USB port that is capable of taking a video signal (speed and memory); see Fig. 3.3.

Telescopes and Their Impact

The next question is, "What telescope to use?" This is a question that comes to mind for many people. The answer is just about any telescope will do, including most Newtonians. However, not all telescopes are created equal.

From Table 3.1 we can see that focal length plays an important role deciding what telescope to use. This is especially true for astro-video cameras that have a 1/3-inch sensor size. The rule of thumb currently used today is that the focal length of a telescope should not be more than approximately 1000 mm for a ½-inch sensor and 800 mm for a 1/3-inch sensor. The large image sizes and small fields of view associated with astro-video cameras make finding and tracking objects in space sometimes a difficult task to do. Selecting a telescope having a relatively short focal length will help mitigate this issue but not eliminate it. Another telescope issue is infrared flaring and chromatic aberration, both of which are issues with refractors.

An accurate GOTO telescope is more than welcome; for most people it is a necessity. The telescope's mount is also part of the equation. If a very sturdy German equatorial GOTO mount with excellent tracking and accurate GOTOs is used, then a telescope having a long focal length is easier to use. But, even then, focal reducers are typically employed to reduce telescope focal length. However, if a lightweight alt-azimuth GOTO mount without precision gearing is used, then fast telescopes with focal lengths of less than 600 mm are advantageous.

Refractors—small aperture, fast, achromatic refractors in particular—have issues with chromatic aberration. This is especially true if they are used for imaging with an astro-video camera. Chromatic aberration can have an impact both on visible light as well as near infrared. Some astro-video cameras sensors, such as the ones used by entry-level cameras, have no infrared filters and are very sensitive to infrared. The same optical principles that cause green light not to focus with red and blue in an achromatic refractor also hold true for infrared and infrared does not focus with red and blue. This inability to focus infrared at the same focal point as blue and red may result in star flaring. The issue is easily resolved with an IR filter.

One filter used for small refractors that many people have is the Baader Semi APO filter. This filter is a Neodymium filter that attenuates some artificial light in the visible spectrum, infrared above 700 nm and blue below 420 nm, with reduced transmission between 460 and 420 nm. The filter is designed to reduce the blue fringe around bright white stars when viewing with an eyepiece. It is not designed for photography and does not correct for chromatic aberration in images produced by a camera. The blue fringe around bright white stars will remain. However, the Semi APO filter does perform well as an IR and UV filter with an astro-video camera and a small refractor and can be used if you have one.

Most amateurs starting to explore video astronomy also plan to use a telescope and mount that they already have. Discussions on Internet groups and forums point to one telescope widely in use as the workhorse of amateur astronomy, an 8-inch Schmidt Cassegrain telescope (SCT) at f/10 having a focal length of 2000 mm. From Table 3.1 we can see that the field of view in an 8-inch SCT is very small and even smaller in larger aperture SCTs. To work around this issue, a 0.63x SCT focal reducer is typically used with Schmidt Cassegrain telescopes to reduce their focal lengths.

Although a 0.63x SCT focal reducer greatly improves the field of view, especially for SCTs having an aperture of 8 inches or less, the resultant field of view is still on the small side. To combat this, many astronomers use a generic 1.25-inch 0.5x focal reducer. Needless to say, a Meade 0.33x SCT focal reducer is a welcome find for many SCT users. Some amateurs stack focal reducers on their SCTs, e.g., a 0.5x and a 0.63x reducer giving a final result of around 0.3x. This works with varying degrees of success.

One astro-video camera manufacturer sells a variable reducer that is adjustable in six steps between 0.63x and 0.29x that fully illuminates a 1/3-inch sensor. For SCTs, a 2-inch 0.18x focal reducer is available for 8-inch and larger SCTs. This unorthodox usage of focal reducers may seem odd to people with a background in astrophotography, as focal reducers typically produce vignetting. This is also true for an astro-video camera. However, vignetting is not as severe as with an astronomical CCD or DSLR camera because the very small sensors used in these cameras are more easily illuminated.

Newtonian telescopes have no issues with chromatic aberration and often have focal ratios of f/5 or faster. This means that even a large aperture Newtonian, say a 10-inch f/4 scope will have a relatively short focal length. Little mention is made on the Internet forums of Newtonians having back focus issues with astro-video cameras. One video camera manufacturer, Astro Video Systems, has a budget-level astro-video camera that fits inside a 2-inch focuser; thus, the camera can come to focus with any Newtonian that has a 2-inch focuser.

Two vital accessories for any telescope are eyepieces and a finder. Although the astro-video camera can eliminate the need for eyepieces, the need for a usable finder remains. Any finder can locate the alignment stars needed to set up a GOTO telescope mount. However, if a tracking mount is used the narrow field of view typically associated with an astro-video camera means that an optical finder is needed to allow star hopping for finding the targeted object. An 8 × 40 finderscope is about as small as is useful for this task.

One very popular method used with video telescopes is to do the GOTO alignment and, if required, the precise polar alignment using eyepieces and then switch to an astro-video camera to find and view objects in the night sky. For this method, a 25-mm eyepiece and a 12.5-mm reticle eyepiece work well for finding and then centering alignment stars. Having said this, some astronomers have the equipment and skills to do their GOTO and polar alignments using their video cameras.

One excellent telescope for beginners to use for video astronomy, especially if they are also fairly new to astronomy, is the ever popular 80-mm, f/5, short tube refractor. Its fast focal ratio and short 400-mm focal length makes it very easy to use for aligning GOTO mounts and then finding and tracking objects in the sky. Several companies sell these little refractors, and some are better than others. Attributes to look for if you decide to start with a short tube refractor are:

- Fully coated optics
- Metal telescope tube
- Metal focuser (a two-speed Crayford is very nice to have but not necessary)
- Vixen dovetail
- Dew shield

Short tube f/5 refractors are also available with apertures of 90 and 102 mm having focal lengths of 450 and 500 mm, respectively. These telescopes are also excellent as starter telescopes for video astronomy. They typically come with a 2-inch focuser. An inexpensive 0.5x focal reducer is an excellent accessory for these short tube refractors. An IR cut filter is also required to prevent infrared flaring.

Telescope Mounts

The large image size and small field of view typical of video telescopes makes having a GOTO mount that can accurately find and track objects a very welcome and, for most people, required component. Although a few people may use tracking mounts, nearly all video astronomers use a GOTO mount to find and track objects with their video telescopes.

Like most astronomy uses involving cameras, the better the stability and the more accurate the tracking of the mount, the better the final image. For video astronomy, as in astrophotography, an equatorial mount has an advantage over an alt-azimuth mount. However, unlike astrophotography, in video astronomy an alt-azimuth mount can compete favorably with an equatorial mount over a very wide range of usage.

One major advantage German equatorial mounts have is that they can adjust for field rotation if precisely polar aligned. This advantage has little impact with a video telescope unless its camera is processing images using its internal stacking capability. For many applications the integration times used to capture an image are too short for field rotation to be noticeable, and an equatorial mount has no advantage over an alt-azimuth mount. Only when a camera's 3D-DNR features are coupled with integrations longer than X256 does field rotation become a significant issue, and the equatorial mount has a real advantage. This is because the camera's stacking programming cannot adjust for field rotation. As the image rotates due to field rotation, photons spill over from one pixel into adjacent ones. This reduces the sharpness of an image and may even cause blurring at high rates of field rotation when an alt-azimuth mount is used. True, you can simply choose to not use the noise reduction features of the camera. Though the resultant image will be noisy, it remains usable for visually observing the night sky.

A high percentage of equatorial mounts used with video telescopes are German equatorial mounts designed for astrophotography. These mounts are large, heavy, and bulky; often weighing well over 55 pounds (25 kg). They provide very stable platforms for telescopes, have very accurate GOTOs, and excellent tracking capabilities. All are available for purchase as a standalone item, although some are also sold bundled with a telescope.

Three German equatorial GOTO mounts that are notable exceptions to the large and heavy characteristics are the SkyWatcher/Orion EQ3Pro, the iOptron SmartEQ PRO, and the iOptron ZEQ25 mounts. Although the SmartEQ PRO and the EQ3 mounts do not offer the stability or payload capabilities of the heavier mounts, both are very suited for a video telescope and both are available for purchase without a bundled telescope. The iOptron ZEQ25 is also very light and portable. It is very suited for traditional astrophotography and has a much heavier payload capacity as well as more accurate GOTO and tracking abilities than either the EQ3Pro or iOptron SmartEQ PRO do. It is also much more expensive and is sold without a telescope bundled as part of the package.

Not to be outdone, alt-azimuth mounts that are very stable and that have excellent GOTO and tracking accuracy are also widely used for video telescopes. Most of these mounts have dual tines and are only sold bundled with an f/10 Schmidt Cassegrain telescope having an aperture of 8 inches (200 mm) or larger. Some currently popular telescopes using this type of mount are the Celestron CPC series and the Meade LX 90/200 series. Like their equatorial counterpart, they are heavy and bulky. Unlike their equatorial counterparts, the optical tube assemblies bundled with these mounts are essentially permanently fixed to the mount. Removal, while possible, is difficult and seldom done. These heavy-duty alt-azimuth mounts are often placed on a wedge that converts them to an equatorial mount and are successfully used for astrophotography. They are not sold without a telescope.

Meade and Celestron offer medium-weight alt-azimuth mounts bundled with Schmidt Cassegrain telescopes that also have accurate GOTOs and tracking (Meade LS series and the Celestron Evolution and SE series). iOptron also offers medium-weight alt-azimuth mounts in their mini-Tower and AZ mount pro series. These mounts are typically used in the alt-azimuth mode, as their designs create some balance issues when used with a wedge. Only the iOptron mounts are offered for sale without being bundled with a telescope.

Many entry-level alt-azimuth mounts are available from Meade, Celestron, SkyWatcher, iOptron, and Orion. None of these mounts is suitable for use with a wedge. The Meade mounts have their OTA permanently attached, and switching OTAs is not easily done. Although the Meade telescopes can be used with an astro-video camera issues exist related to camera clearance and mount stability. The Celestron, Skywatcher, and Orion mounts are very similar in design and all are manufactured by the same company in China. iOptron has its Cube series that has several robust and accurate entry-level alt-azimuth mounts. Only iOptron, SkyWatcher, and Orion offer their entry-level mounts on the market without being bundled with a telescope. Although these mounts do have issues with vibration and tracking, they are suitable for video telescopes having short focal lengths and make excellent lightweight and portable video telescopes for even experienced video astronomers.

The only real limitation associated with alt-azimuth mounts from a video perspective is they cannot compensate for field rotation. The impact of field rotation with an alt-azimuth mount is dependent upon the position of the object in the sky and the location of the telescope on Earth. See the field rotation tables in Appendix B in this book. In some areas of the night sky, the rate of field rotation is very small, and its impact will not be noticeable even if the camera's 3D-DNR feature is used with long integrations. However, field rotation rates in the area of the sky near the zenith are sufficient to produce noticeable degradation of the final image in a matter of a few seconds.

All of the major telescope manufacturers offer portable telescopes on a lightweight GOTO alt-azimuth mount at the lower end of their pricing structure. These telescopes are often referred to as entry-level telescopes, and mounts on these telescopes are typically single arm with a dovetail saddle for placing on the telescopes. The telescopes range from small 80-mm refractors to 150-mm SCTs, MAKs, and Newtonians. In between are 90-, 102-, and 125-mm Cassegrain telescopes as well as 114-, 130-, and 150-mm Newtonians. Prices start at $300 and climb upwards towards a $1000 for a 150-mm MAK or $1200 for a 200-mm SCT.

Celestron offers four series of telescopes in this category, their Nexstar LMC, SLT, Prodigy, and SE telescopes. Meade has its ETX and StarNavigator (DS2000) series. iOptron offers three small aperture telescopes on its entry-level cube mounts, while SkyWatcher has its SynScan AZ GOTO telescopes. Perhaps the most complete line of entry-level telescopes is offered by Orion with its Star Seeker IV series of telescopes. These telescopes and their mounts do very well for visual work with an eyepiece. Although none is designed for photography or video astronomy, people are very successful using them for video astronomy as well as for very short exposure astrophotography.

A very large number of people fairly new to astronomy purchase one of these entry-level telescopes as their first telescope due to their ease of use and price. Experienced amateurs sometimes buy them as a "grab and go" telescope that is easy to transport and fast to setup and use. These telescopes and mounts all have some characteristics in common:

- They have moderate to fairly accurate GOTOs.
- Their tracking accuracy is fair to excellent for visual work but marginal for astrophotography.
- Their lightweight tripods are susceptible to vibration
- They are very easy to transport.
- They come bundled with small to medium aperture, lightweight telescopes.
- Other than the Orion, iOptron and SkyWatcher mounts, purchasing a mount without a bundled telescope often requires some effort and time to locate.

These entry-level telescopes are lightweight and very portable. They all have small to medium apertures and newcomers to astronomy typically soon want to see deeper in space than most of these scopes can go and purchase a larger telescope. An astro-video camera offers an excellent and much less expensive alternative to buying a new telescope and mount. Like the more expensive alt-azimuth mounts bundled with SCTs, entry-level alt-azimuth mounts can produce a useful image for exposures of 20–30 seconds without any major influence from field rotation for most areas of the night sky.

However, the same is not true for their GOTO accuracy and tracking movements. These mounts typically have sufficient accuracy to put an object somewhere within the field of view using a 25- to 32-mm eyepiece with the telescopes that are typically bundled with them. Using a 0.5x focal reducer improves the GOTO accuracy even more. With a Schmidt or Maksutov Cassegrain telescope, the object often is not visible, and a spiral search is needed to locate the targeted object. This limits camera integrations to about X128 maximum if a search is required (~2-second delay). Once an object is located then the integration is adjusted to produce the desired image on the TV monitor or computer screen.

Small aperture refractor telescopes having short focal lengths of 500 mm or less are bundled with entry-level mounts. With a 0.5x focal reducer, these mount and telescope packages will typically put an object near the center of the field of view of an entry-level astro-video camera.

Although an object appears stationary to the human eye viewing through an eyepiece for telescopes using budget-priced mounts, the object is actually slowly drifting, due to errors in tracking movements. Although not noticeable to the eye, these movements are significant to a camera. Typically this movement is small. Over the 21 seconds needed by an entry-level astro-video camera to integrate up to 256 frames with the camera's 3D-DNR feature selected, the movement may soften the image slightly, but the camera still produces an acceptable image. By soften we mean very slight blurring of the image, making details not as sharp and distinct as they could be.

For an image made using the 3D-DNR feature of an astro-video camera, the tracking movement during integrations of 526 frames or more is often significant and can have a similar impact upon the final image, as field rotation does. The image produced can have bloated stars and will not be sharp. Tracking movement is not constant but seems to vary among mounts and is also dependent somewhat upon where in the sky the object is located. Also, some entry-level mounts, due to luck with manufacturing tolerances, are more accurate and track better than their peers even though the mounts are identical.

GOTO accuracy and tracking precision are often improved by selecting alignment stars separated by at least 60 degrees and 30 degrees above the horizon. Also, if possible, select one of the alignment stars near the object(s) selected for viewing that night. Accurately leveling the mount may also help reduce tracking movements.

The designs of these budget mounts vary among the different manufacturers. Some mounts have higher levels of accuracy than other mounts; some are more sturdy, for example. All of the mounts work very well for their designed purpose of observing the night sky using an eyepiece, but some mounts such as the Meade DS2000 (Star Navigator) and the Celestron LCM have bigger GOTO and tracking issues with an astro-video camera than do the other mounts. Table 3.2 is a listing of mounts bundled with entry-level telescopes along with a subjective assessment of their relative suitability for use as a video telescope mount.

Table 3.2 Entry-level alt-azimuth mount suitability for astro-video usage

Manufacturer	Mount	GOTO	Tracking	Vibration	Flexibility	Score
Celestron	4/5 SE	11	11	11	11	44
Orion	StarSeeker IV	11	11	11	10	43
iOptron	Cube A	7	7	7	9	30
iOptron	Cube G	7	7	7	8	29
SkyWatcher	SynScan AZ	7	7	7	7	28
Celestron	Prodigy	7	7	7	6	27
Celestron	SLT	7	7	7	5	26
Meade	ETX90	7	7	7	4	25
iOptron	Cube E	3	3	3	3	12
Celestron	LCM	2	2	1	2	7
Meade	DS2000	1	1	2	1	5

Note A score of 11 in a category does not mean that the mount is perfect, only that it is perceived as the best of the mounts evaluated for that particular category. Conversely, a score of 1 does not mean that mount is defective, only that it is ranked lowest in a particular category. Categories having identical scores are equal.

Does this mean that an entry-level alt-azimuth mount cannot be used for video astronomy? No, not by any means. All it means is that the final image produced by budget-priced alt-azimuth mount may not have the sharpness and details that more expensive alt-azimuth mounts or German equatorial mounts are capable of producing. There is also a workaround for this problem if the camera is being used for astrophotography. An external stacking program that can adjust for both field rotation and mount tracking movements can be used to produce a final image that is competitive with the images made with more expensive mounts.

Budget Entries into Video Astronomy

Unlike conventional astrophotography, video astronomy does not require a major investment in mounts and other equipment. One inexpensive telescope and mount combination is shown in Fig. 3.4. There are many others.

Here are some suggested telescope and mount combinations that perform well for video astronomy:

One excellent telescope and mount combination is the Orion ST80A 80-mm f/5 refractor on an iOptron SmartEQ PRO mount. The German equatorial mount and telescope combination cost $710. The payload of the mount is more than adequate

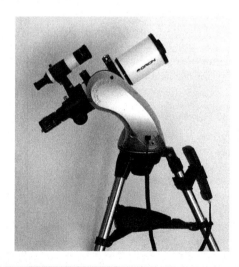

Fig. 3.4 Budget video telescope

for the ST80A with a video camera and other accessories attached. The optics of the 80-mm f/5 refractor work very well for video astronomy. The 400-mm focal length of the ST80A without a focal reducer and 200 mm with a 0.5 focal reducer makes locating objects easy to do while at the same time producing an image size sufficiently large enough to see details of most, but not all, objects in the night sky.

The Orion StarSeeker IV mount is sold as a standalone unit for $380. Adding the $210 Orion ST80A to the mount makes a very economical but capable telescope and mount combination for video astronomy. The StarSeeker IV mount has very good GOTO and acceptable tracking accuracies. The mount has a very solid tripod and low levels of vibration. It can be manually moved without losing its alignment. As previously discussed, the ST80 is an excellent telescope for video astronomy.

Another combination that works very well is the ST80A on a Celestron Nexstar 4/5 SE mount. The 4/5 SE mount is very stable, has an extensive database, and has a built-in wedge that can easily convert the mount into an equatorial mount. Celestron's All Star Polar Align features makes obtaining a polar alignment quick and easy to do. Finding a 4/5 SE mount can be difficult, though, as they are no longer stocked by Celestron as a standalone mount. However, with a 0.5x focal reducer, the $500 4 SE or the $700 5SE have focal lengths of 660 or 635 mm, respectively; either should do well for video astronomy.

In Europe, one excellent option is the SkyWatcher SynScan AZ GOTO mount with a SkyWatcher StarTravel 80-mm f/5 refractor. This combination does not come as a bundle, so the components must be purchased separately. The SkyWatcher SynScan AZ GOTO mount is very similar to the Celestron SLT mount, but, unlike the SLT mount, it has a very large database of objects in its memory. The mount is very accurate as far as GOTOs are concerned. The same is not true for tracking. Although the mount keeps an object in view of an eyepiece, it hunts around considerably. This movement is not visible to the eye but is to the camera. The results, while acceptable, are a soft image with some bloated stars. The StarTravel 80-mm f/5 refractor is identical to the Orion ST80A. The telescope and mount combination costs 440 Euros in Europe. A version bundling a 102-mm f/5 refractor with the SynScan AZ GOTO mount is available for 435 Euros if greater light collection and slightly longer focal length are desired.

Another excellent combination in Europe is the SkyWatcher All View mount in combination with the SkyWatcher StarTravel 80. The combination costs 540 Euros. The All View mount is an alt-azimuth designed specifically for photography. It has excellent GOTO and tracking accuracy as well as a solid tripod to provide vibration-free action.

At $290, the Meade ETX 80 is perhaps the lowest priced telescope that is usable for video astronomy. It is an 80-mm f/5 short tube refractor having a focal length of 400 mm without a focal reducer and 200 mm with a 0.5 focal reducer. The mount has an extensive database of deep space objects, and it should handle the weight of an astro-video camera if balanced. The lightweight tripod will cause vibration issues, especially under windy conditions. In contrast the $320 ETX 90 is a Maksutov Cassegrain telescope having a focal ratio of 13.8 and a focal length of 1250 mm without a focal reducer and 625 mm with one. Its tripod is very sturdy and has a built-in wedge for operation in the equatorial mode. A 0.5x focal reducer will definitely be needed.

Among the lowest-priced bundled telescopes and mounts suitable for video astronomy is the $350 iOptron Cube E R80. The Cube E R80 is the iOptron Cube E mount bundled with iOptron's 80-mm f/5 short tube refractor. The Cube E mount at $270 is the least expensive mount in the iOptron lineup. Its major negative is its rather small database that, although adequate for work at the eyepiece, is a bit spartan for a video telescope. In addition to the limited data base of the Cube E mount, the refractor telescope bundled with the mount uses a plastic focuser that may have issues with focusing. The mount has an upgrade to the Cube G mount and is sold as the Cube G R80. The $450 Cube G R80 uses the same telescope as the Cube E R80, and thus has the same focuser issues. However, the Cube G mount has an extensive database as well as GPS positioning and other features that make it more suitable for video astronomy.

There are two other mounts in the Cube series, the $420 Cube A and the $428 Cube Pro. Although both are excellent mounts, the $500 SmartEQ PRO or the $380 Orion StarSeeker IV are the mounts to select at this price level.

The $425 Celestron Nexstar 102 SLT is an alternative to the iOptron mount and telescope bundles. It is a 102-mm f/6.5 refractor having a focal length of 660 mm without a focal reducer and 330 mm with a 0.5x focal reducer. It is very suitable for video astronomy and for visual work using an eyepiece as well. The $420 Celestron Nexstar 127 SLT is a 127-mm f/12 Maksutov Cassegrain telescope having a focal length of 1500 mm without a focal reducer and 750 mm with a 0.5x focal reducer. Although usable as a video telescope, its relatively long focal length, with or without a focal reducer, coupled with the GOTO accuracy of the STL mount, will create issues acquiring an image in the field of view of the camera. The smaller $325 Celestron Nexstar 90 SLT is also a Maksutov Cassegrain telescope with a focal ratio of f/13.3. It has a focal length of 1200 mm without a focal reducer and 600 mm with a 0.5x focal reducer. As with the 127 SLT, a focal reducer will definitely be needed when using an astro-video camera with this telescope. The database in the SLT mount has 4000 objects. This is very adequate for visual work with the eyepiece, but it is on the light side for a video telescope, as the effective apertures of the telescopes bundled with the mounts are essentially tripled, e.g., a 102-mm refractor has the seeing ability of a 306-mm telescope.

As you can see, a large variety of telescope and mount combinations are available for video astronomy. Which one is best really depends upon your budget and requirements. Unlike traditional astrophotography, where the mount is all-important, for video astronomy the telescope also plays a major role. Keep in mind the impact of a telescope's focal length as you decide which telescope you want to use. From a cost and performance perspective an 80-mm f/5 mm refractor on a mount such as the Orion SkySeeker IV is an excellent combination for an entry-level astro-video camera.

Chapter 4

Light Pollution and Filters

What Causes Light Pollution?

Simply stated, light pollution is any light that is not wanted and typically refers to the large, artificially produced light glow in the sky that seems to exist wherever humans live in any significant numbers. The extent of the phenomenon is worldwide, as shown by NASA's famous photograph (Fig. 4.1).

A glance at NASA's Earth Lights photograph shows the light trails of many countries, continents, and surface transportation corridors made by our streetlights, car headlights, security lights, advertising, and other lighting that are all so prevalent with our modern urban society. With very few exceptions, much of this light is directed upward into space, doing no one any good on Earth below. The photograph shows that the bright areas are associated with urbanization, not population. Europe, the United States, India, and China glow as equals, as does Egypt from the Aswan Dam to the Mediterranean Sea and Japan in the Pacific Ocean. Northern Canada, Africa, Australia, Antarctica, and, to a lesser extent, South America remain dark.

Based on statistics from the World Health Organization, at the end of the year 2014, 54% of the global population lived in urban areas. The urbanization of the planet is expected to continue. By the year 2050, 70% of Earth's population is expected to reside in urban areas. Already countries in North America and Europe have as much as 80% of their populations residing in urban areas.

How does all this fit into astronomy? First, most amateur astronomers reside in developed countries having a high standard of living, as impoverished people cannot afford telescopes. Second, the percentage of amateur astronomers living in light-polluted areas is pretty much the same as the percentage of the general

Fig. 4.1 Earth Lights (Image courtesy of NASA/GSFC)

population. After all, amateur astronomers, like everyone else, must live where the jobs are, and most jobs today are in urban areas.

If these two observations are true, then urbanization and its companion light pollution cannot be expected to go away. Advances in lighting technology may reduce the magnitude of light pollution in the future, but it is reasonable to assume that most amateur astronomers now and in the future must learn how to cope with our light-polluted night sky. This is true both for visual work and for cameras, whether the camera be a DSLR, an astronomical CCD camera, or an astro-video camera.

What exactly is light pollution? Simply put, light pollution is any unwanted light. It has two components, trespass light and sky glow. From an astronomical perspective, trespass light is light shining directly onto an observing site, such as a nearby security light or a streetlight. Earth's atmosphere is never truly dark but glows at night. This "sky glow" has two components—natural sky glow created by nature and artificial sky glow created by humans.

Why is light pollution a concern? Like the blinding bright lights of an inconsiderate driver in an approaching automobile at night reduces our ability to see where we are going, light pollution reduces our ability to see objects in the night sky. Since light pollution is not likely going away anytime soon, astronomers must find ways to mitigate its impact. For this task astronomers have three major weapons: light screens, light pollution filters, and video telescopes. None of these tools match a dark sky, but they do allow amateur astronomy to continue in city skies.

Some urban astronomers will have the luxury of viewing from private observatories or a dark spot in their private gardens and yards. Unfortunately for many this is not an option, especially for people who live in apartment complexes or condominiums, among tall buildings. They must travel away from their homes to either a nearby location such as a park or out of the city to darker suburban or rural skies. Transporting a large telescope kit requiring multiple trips is often intimidating and generally not that practical. In this case a truly portable observatory, one that can be easily carried aboard public transportation such as a subway, can be the difference between viewing the night sky and joining the ranks of armchair astronomers. A video telescope with a small aperture, fast refractor on a lightweight alt-azimuth mount is a good candidate for a portable observatory.

Making the decision to get out of the armchair and start observing is the first step many must take to combat our light polluted night sky. For the folks that do and who choose to use a video telescope three surprises are in store:

- The stars are still in the sky, and with a telescope you can see them in an eyepiece along with many deep space objects.
- Replace the eyepiece with an astro-video camera, and objects having a surface brightness two to three magnitudes less than sky glow are not only visible but are visible in color.
- Add a frame grabber to digitize and store the analog television signal in a computer, and you can produce photographs of the object you are viewing.

Trespass Light and How to Mitigate It

Earlier, light pollution was divided into two categories, "trespass light" and "sky glow." Trespass light, often called stray light, is the easier of the two categories of light pollution to combat, at least for video astronomy. A video telescope is viewed on a TV or computer monitor that may be located inside the comfort of your home, so why even worry about trespass light and its companion, glare? That's a good question. If the trespass light comes from some direction not available for viewing because of an obstruction such as a tree or a structure and blocks access to the night sky, it may be more of an annoyance than a problem for a video telescope. In this case a hood or a fabric cover may be all that is necessary to set up the telescope or to find an object to view (see Fig. 4.2).

Keep in mind that trespass light can also enter a telescope, especially if it comes from a direction in proximity to the object you want to view. This has negative consequences. Halos and reflections of the light can appear on images. The glare from trespass light, if bright enough, can reduce the contrast and details of the object we want to view or, in the worst case, can even make the object not visible at all.

A good observing site will allow viewing a large area of the night sky. Often one or more sources of trespass light originate from directions where you want to point your telescope. Typically the sources of this trespass light are streetlights, parking lot lights, and security lights. One particularly annoying security light source might be the next door neighbor. If your neighbor has an offending security light, invite him over to view the night sky with you and your telescope. A cold drink on a hot summer night spent with your neighbor and your telescope often solves the security

Fig. 4.2 Temporary stray light shield

light problem, but then there are folks who simply fear the dark and light up the world regardless of others' needs.

Convincing owners of street and parking lot lights, government or private, to turn them off is difficult, especially in litigious countries. Even if you are successful the lights will typically stay on until midnight. This will cost you half the night. The best approach with street and parking lot lights is a long range strategy to convince the city or a parking lot owner to replace the offending lights with energy efficient light fixtures that direct light only to where it is wanted. This approach is also economical; the energy saved will more than pay for the cost of the efficient light fixtures. The likelihood that this strategy will succeed is high enough to warrant pursuit but will also take a considerable investment of your personal time and effort. In all likelihood, if you have a trespass light situation, you will need to take your own measures to mitigate the problem. For this, light screens are the major weapon available for your use.

Keep in mind that light screens come in many forms and need not be elaborate or permanent. Some possibilities are listed below, but many others exist:

- a blanket, tarp, etc., thrown across a rope stretched between two vertical supports (such as a small tree, a utility shed and your house, clothes line);
- a free-standing, portable frame made of wood or plastic piping with fabric or plastic sheeting;

- a blanket, tablecloth, tarp, etc., thrown over a wheeled garment rack like the kind used in hotels;
- a permanent fixture disguised as a trellis for your garden or screen for your firewood pile;
- a hedgerow trimmed to the proper height (choose a plant that has thick foliage year round);
- the shade provided by a utility shed or a parked vehicle such as a van;
- a large oversized hooded jacket or large thick beach towel to drape over your head and the eyepiece/camera on your telescope;
- an eye patch;
- an observatory of your own.

Do a survey of your property or the site where you plan to observe and identify possible observing locations that offer good windows into the night sky. Many astronomers in the northern hemisphere like to at least have a clear view to the southeast, south, and southwest with a view of Polaris to the north. The initial step of the survey is best done during the day so you can make notes of the terrain, access to electric power if desired, the presence of lawn sprinklers, and any other positive or negative attributes of the location. Lawn sprinklers are frequently a "show stopper" if you do not personally control their hours of operation.

At night go out and check each location for trespass light and make notes where the light originates and how (location, size, and orientation) light screens can mitigate light from each source. Keep in mind that if you must travel to a location away from home, your options for using light shields are greatly reduced, especially if you plan to backpack, use public transportation, or must move everything in one trip.

The point here is that if you have a location that is open to the night sky, with a little imagination and ingenuity you can generally shield trespass light and have a usable observing or photography session. This is especially true for a video telescope with a TV monitor for viewing the night sky. As you contemplate your options, don't forget that many neighborhoods, especially in the United States, have homeowner associations with their long list of things that are acceptable and things that are not. Here, portable shields that you assemble each night and remove at the end of an observing session are generally required; however, sometimes you can disguise a permanent light shield as a garden trellis, etc.

In addition to using light shields to screen trespass light from falling onto the observing location; a light shield or dew shield on the telescope itself is often very effective as long as the telescope is not pointed directly into the vicinity of the stray light. If you don't have a light shield or dew shield on your telescope, you can make one. All that is needed is cardboard and tape if your telescope is not too big. A good length for a light shield is equal to the diameter of the telescope tube. A light shield is particularly helpful for many Newtonians to prevent trespass light from striking the secondary mirror or possibly from entering through the focuser drawtube.

In many parts of Europe and other areas of the world, housing structures are constructed of reinforced concrete and frequently have flat roofs. Although not ideal for several reasons (turbulence, vibration, etc.), the flat roofs and balconies of these structures due to their height often have little or no trespass light issues and, for many, are the only areas available to set up at night. If you attempt to use a balcony or rooftop, keep in mind that while the near real-time view on your TV monitor may appear "rock solid" and vibration free, the camera may see a different picture if you are using it for imaging.

Even with light shields, trespass light and glare can play havoc with night vision. Light screens work only while you stay in the shade they provide. They only offer partial protection, as plenty of light is reflected by other surroundings into the observing location. It truly never gets dark even in the shade of a light shield. Fortunately, this is not as much a negative for video astronomy as it is for visual observing or traditional astrophotography. With a video telescope, the TV monitor effectively prevents dark adaption but also its display is bright enough where night vision is not needed.

Although adapted night vision is not needed for viewing a TV monitor, it is helpful for using the telescope's finder scope, aligning the mount, composing an image, and focusing. During the cooler months, an oversized hood works well to further shield stray light and also helps keep you warm. During the warm part of the year, a lightweight piece of fabric such as dark colored beach towel draped over your head works well while you are at the eyepiece or trying to use the camera to find an object, focus, etc. If you must leave an area protected by a light shield, red goggles or an eye patch can help retain your night vision if that is important to you.

Natural Sky Glow

The surface of Earth is never truly dark even in the darkest of places. Earth's atmosphere literally glows at night. This sky glow has two components: natural sky glow and artificial sky glow. Natural sky glow covers Earth with its heavenly light. Artificial sky glow hovers in the sky wherever significant numbers of humans are present.

Natural sky glow is created by nature and is present globally. It has several sources. Everyone is familiar with the Moon and the light the Moon shines on Earth when it is present. Moonlight is light reflected directly from the Sun or indirectly reflected off Earth to the Moon and back to Earth again. When the Moon is visible, moonlight typically dominates natural sky glow, especially between the first and third quarter.

Another significant source of natural sky glow is the excitation of oxygen atoms in the upper atmosphere by particles shooting out from the Sun. The entire Earth is enveloped by a green glow visible to our astronauts in space and under certain conditions by observers here on Earth. As expected the glowing oxygen is more pronounced just after sunset and just before sunrise but is present all night long. This glow should not be confused with auroras, which are colorful displays of ionized gases in the far northern and southern areas of our planet.

Other sources of natural sky glow are light from stars, galaxies, planets, *Gegenschein*, and zodiacal light. Gegenschein is a faint oval glow of light in the night sky about 20 degrees across opposite the Sun. It is caused by sunlight that is backscattered by interplanetary dust orbiting the Sun in the ecliptic plane. Zodiacal light is a diffuse triangular pattern of white light in the night sky extending from the Sun along the zodiac and is best seen in the fall and spring. It is sunlight scattered by space dust.

The Nature of Artificial Sky Glow

This is created by humans. As shown in Fig. 4.3, it is the glowing light dome seen over populated areas with the magnitude of the glow proportional to the size of the metropolitan area. The source of the glow is light directed straight upward into the sky and light reflected indirectly from the ground and then up into the sky. This reflected light is scattered by humidity, clouds, and airborne aerosols (particulate matter such as smoke, haze, dust, and particulate air pollutants), sending a good portion back toward the ground that renders all but the Moon and the brightest of stars and planets invisible to the human eye. Particulate and aerosol levels vary from night to night, which in turn varies the amount of light scattered back to Earth.

The light produced by humans that is directed into the sky is not homogenous. It is a mixture, a "stew" of light from several technologies, the major ones being:

- incandescent
- florescent

Fig. 4.3 Artificial sky glow over Athens

- mercury vapor
- low pressure sodium
- high pressure sodium
- LED
- colored advertising (argon, neon)
- halogen
- xenon

This mixture of technologies produces a light glow that covers the span of light frequencies visible to the human eye and, as discussed later, complicates efforts to mitigate the impact of artificial sky glow. The major contributing sources to artificial sky glow are streetlights, security lighting, advertisement lighting, and automobile headlights.

Until recently, the dominant lighting technology used for street and security lighting was Mercury vapor and sodium lights, which are narrowband and can be filtered. However, conversions to LED lights are rapidly making inroads as cities seek to find ways to cut operating costs. Unfortunately, these conversions to LED technology are not always properly done, and while energy consumption may decrease, sky glow increases.

Automobile headlights are a major contributor in the early to mid evening hours. Halogen headlights dominate automotive lighting followed by xenon headlights, and with LED headlights currently entering the market.

Recall the magnitude scale used to denote the brightness of stars. That brightness, of objects such as nebulae and galaxies, is measured using the same magnitude scale used for stars but spread over the surface area of the object and is expressed in units of magnitude per square arc second. The brightness of sky glow is measured for each square arc second of the night sky and also uses the same magnitude scale. In other words the scales used to measure sky glow and deep space objects such as nebula and galaxies are the same. (*Note:* If you need more information about magnitude and brightness, see Appendix A in this book.)

Natural sky glow with a full Moon has a magnitude of about 18 per square arc second. In comparison, the brightness of artificial sky glow in the middle of one of Earth's mega cities such as New York or London typically has a magnitude of 17 per square arc second. Natural sky glow at a true dark location with no artificial light glow or moonlight has a magnitude of 22 per square arc second.

To an astro-video camera, or most any other camera, sky glow is the same as the photons from some distant object. In other words, a camera cannot tell whether photons come from a distant deep space object or from sky glow. At first glance one might conclude that an object with a surface brightness equal to or less than sky glow is not visible and cannot be imaged. However this is not necessarily the case.

The surface brightness of an object is an average over its entire surface. In reality, the object's brightness varies across its surface, with the brighter areas of the object having a magnitude brighter than the average value and dimmer areas having a lower than average value. Thus, when an object's surface brightness equals the magnitude of sky glow, some areas of the object, such as the bright core of a galaxy, are brighter than sky glow and remain visible in the eyepiece while the rest of the object, such as a galaxy's spiral arms, cannot be seen with the human eye because the contrast may be too low or that part of the image is simply buried in sky glow. However, a camera is different from the human eye and can integrate an exposure over time. This capability can produce images having details not visible to the human eye, even though the camera's exposure time is very short.

How Astro-Video Cameras Pierce Artificial Sky Glow

One aspect of astro-video cameras is the ability to view objects in near real time that are buried in about two magnitudes of the local sky glow. Recall that signal increases linearly with time, and photon noise increases as the square root of time. Another way of stating this is that noise increases with the square root of the number of photons collected. Also, keep in mind that the camera cannot differentiate between photons emitted by sky glow or photons emitted by a deep space object. This produces an interesting problem, as we now have two sources for signals, sky glow that we don't want and the deep space object that we do want.

Let's look at how sky glow impacts an astro-video camera viewing an emissions nebula, for an example. For this exercise we will use a 100-mm f/5 refractor and a 1/3 inch astro-video camera. Assume that we are located downtown in a major city. Further assume that the magnitude of sky glow is 17 per sq. arc-sec with a photon flux per sq. arc-sec of 120 photons per second. In the sky is a nebula we wish to view with our video telescope. Its average surface brightness has a magnitude of 19 per sq arc-sec with a peak brightness of 18 magnitudes per sq. arc-sec and a minimum brightness of magnitude 19.5 per sq. arc-sec. Since photon flux is expressed in photons per second, let's assume an exposure time of one second.

Recall that a difference of half a magnitude in brightness is a change of about 1.6 times, one magnitude is about 2.5 times, and a difference of two magnitudes is a change of approximately 6 times. Also recall that surface magnitude is expressed in units of magnitude per square arc-seconds (mag/sq. arc-sec), and photon flux is expressed in photons per square arc-second.

In our example, a difference of 2 magnitudes exists between sky glow and the average surface brightness of the nebula. This means that the nebula's photon flux will be one-sixth that of sky glow, or 20 photons per second per sq. arc-sec. The nebula's peak surface magnitude is one magnitude below sky glow, so the peak photon flux per sq. arc-seconds from the nebula will be 48 photons per second—2.5 times less than sky glow. Sky glow and the minimum surface brightness of the nebula have a difference of 3 magnitudes; thus, the minimum photon flux from the nebula will be 8 photons per second per sq. arc-sec. See Table 4.1 for a synopsis of these values.

Consider the light flux from the sky glow. This flux is 120 photons per second and is essentially uniform. The result is that each and every pixel of the camera's

Table 4.1 Example values

	Surface brightness	Photon flux
Sky glow	Magnitude 17 per sq. arc-sec	120 photons per second per sq. arc-sec
Nebula surface brightness	Magnitude 19 per sq. arc-sec	20 photons per second per sq. arc-sec
Peak nebula surface brightness	Magnitude 18 per sq. arc-sec	48 photons per second per sq. arc-sec
Minimum nebula surface brightness	Magnitude 20 per sq. arc-sec	8 photons per second per sq. arc-sec

sensor records 120 photons per second from the sky glow. Now, consider the photons coming from our distant nebula. They, too, enter the telescope, strike the sensor of our camera, and are added to the photon count of each pixel.

Unlike with sky glow, not all pixels receive the same number of photons. The image emits photons in relation to the brightness of its different areas. The brightest areas emit more photons per second than do the dimmer areas. Thus, pixels exposed to bright areas receive more photons than pixels exposed to dimmer areas. In our example we are using only three points (peak brightness, surface (average) brightness, and minimum brightness). In reality, an image has these plus everything in between.

Recall that in our example the brightest area of the nebula emits a maximum of 48 photons per second. Over the period of one second, pixels exposed to the brightest areas of the nebula will collect 120 photons from sky glow and 48 photons from the nebula. Thus, the pixels recording the brightest part of the nebula will collect a signal totaling 168 photons per second.

Now consider photon noise, which is the square root of the total signal, or 13 photons per second in this case. This gives us a signal-to-noise ratio of 13, which is very high. The problem is that the signal has two sources. One source is sky glow. Sky glow is a source that we don't want and is also our major source of photon noise. The maximum signal from our nebula is 48 photons per second, which is almost 4 times greater than photon noise and 40% above sky glow; thus, it should be easily visible in our image.

The impact of unwanted sky glow gives the impression that the image has an excellent signal-to-noise ratio. However if we look at the situation from another perspective, a different story emerges. For the same reason the camera cannot discriminate between a signal from our nebula and a signal from sky glow, it cannot discriminate between photon noise from sky glow or from our nebula. For the brightest areas of the nebula, photon noise from sky glow and the nebula is equal to 13 photons per second. The actual signal we want, photons from the nebula, is equal to 48 photons per second. This gives us an effective signal-to-noise ratio of 3:7.

Note For the purposes of this book the *effective* signal-to-noise ratio (ESNR) is defined as the "wanted signal" divided by the photon noise generated by all signal sources, "wanted or unwanted."

Using the same logic, pixels receiving photons from areas of the nebula having an average brightness collect a total of 140 photons per second (120 from sky glow and 20 from the nebula). Photon noise is approximately 12 photons per second. This gives an effective signal-to-noise ratio of around 1:7. Areas of the nebula having an average brightness should be visible but will not have much contrast against sky glow and will show considerable noise.

Pixels receiving photons from the dimmest areas of the nebula collect a total of 128 photons per second (120 from sky glow and 8 from the nebula). Photon noise is approximately 11 photons per second, which is greater than the 8 photons per second from the nebula. This gives an effective signal-to-noise ratio of around 0:7. The dimmest areas of the nebula are not visible, as they are buried in noise and have essentially no contrast against sky glow.

The typical entry-level astro-video camera is capable of exposures as short as one-hundred thousandth of a second to integrations as long as 17 seconds (NTSC) or 20 seconds (PAL). With 3D-DNR stacking, a total integrated exposure time of 85 seconds is available for an NTSC standard camera or 100 seconds for a PAL standard camera. With this in mind, let's take another look at our hypothetical nebula to see the impact of increasing the exposure time from one second to 20 seconds.

Recall that in our example the brightest area of the nebula emits a maximum of 48 photons per second. Over the period of 20 seconds, pixels exposed to the brightest areas of the nebula will collect 2400 photons from sky glow and 960 photons from the nebula. Thus, the pixels recording the brightest part of the nebula will collect a signal totaling 3360 photons in 20 seconds.

Now consider photon noise, which is the square root of the total signal, or 58 photons per second in this case. This gives us a signal-to-noise ratio of 58, which is very high. As discussed earlier, the problem is that the signal has two sources. One is sky glow. Sky glow is a source that we don't want and is also our major source of photon noise. The maximum signal from our nebula is 960 photons in 20 seconds which is almost 16.5 times greater than photon noise and 40% above the sky glow; thus, the brightest areas of the nebula should be easily visible in our image. As with the one-second exposure, most of the photon noise contribution is generated by sky glow. Using the concept of the effective signal-to-noise ratio, the signal-to-noise ratio is 16.6.

Using the same logic, pixels receiving photons from areas of the nebula having an average brightness collect a total of 2800 photons per 20 seconds (400 from the nebula and 2400 from sky glow). Photon noise is approximately 53 photons over 20 seconds. This gives a signal-to-noise ratio around 53. The signal from the nebula is 400 photons over 20 seconds, which is almost 4 times the photon noise and about 17% greater than sky glow. The effective signal-to-noise ratio for the average signal is 7.5. Areas of the nebula having an average brightness should be visible but will not have much contrast against sky glow and will show noise.

Pixels receiving photons from the dimmest part of the nebula will collect a total of 2560 photons over 20 seconds (2400+160). Photon noise is approximately 51

Table 4.2 Effective signal-to-noise ratio versus exposure time comparison

	One-second exposure		Twenty-second exposure	
	Sky glow contrast	ESNR	Sky glow contrast	ESNR
Peak surface brightness	40%	3.7	40%	16.6
Average surface brightness	17%	1.7	17%	7.5
Minimum surface brightness	7%	0.7	7%	3.1

photons over 20 seconds. This gives an effective signal-to-noise ratio of 3.1. The total photons detected will be only 7% greater than the number of photons from sky glow. The dimmest areas of the nebula may be visible, but if so, they will have significant noise and little contrast against sky glow.

So how good is the 20-second image? It will show good contrast and details with little noise for the brighter parts of the nebula but will suffer with noise and poor contrast for other areas of the nebula, with the dimmest areas of the image being very noisy with poor contrast. Actually, considering the conditions, the results produced for our hypothetical nebula are noisy but not really that bad, especially considering that the average brightness of the signal from the nebula was two magnitudes less than the sky glow signal. This tells us that some meaningful video astronomy is possible in the worst of skies, provided stray light and glare are controlled.

Table 4.2 is a synopsis of the two examples (one second and 20 seconds). Two attributes are obvious. Increasing exposure time has no impact upon contrast but does improve the effective signal-to-noise ratio of the image.

Note Recall that the values used for photon flux are hypothetical values.

The typical entry-level astro-video camera can stack up to five integrations internally when the camera's 3D-DNR feature is selected. When doing this, the camera can also be programmed to do dark frame subtraction (DPC set to "on") to get rid of hot pixels. DPC increases the process time significantly. The camera's stacking routine stacks pixels; it has limited abilities to align features such as stars, so a polar aligned equatorial mount is required if dark frame subtraction is used. Without dark frame subtraction, the 3D-DNR feature can provide a total exposure time of up to 85 seconds (NTSC) or 100 seconds (PAL).

As we demonstrated, increasing exposure times reduces noise. However, most astro-video cameras are not capable of long exposures and do not have a bulb setting. Here again, the potential improvement has limits. Since the camera cannot differentiate between photons from sky glow and photons from the object; both sky glow and the object's signal are increased proportionally by the same amount.

Fig. 4.4 Light pollution filters

Light Pollution Reduction Filters and Astro-Video Cameras

Light pollution reduction (LPR) filters in some circumstances can reduce photons from sky glow without significantly reducing the photons from the objects being viewed or imaged with a video telescope. An astro-video camera is unique in that it is used as a live viewing device for deep space objects as well as an astronomical camera. It can also operate as either a color imager or a monochrome imager. This dual-use characteristic means that a wide range of filtering technologies is useful for video astronomy.

The term light pollution reduction (LPR) filter covers a wide range of filters, from mild filters passing broad bands of light to strong narrowband filters passing light in bands only a few nanometers wide. To add a little to the confusion about light pollution reduction filters, they are also called nebula filters. The two terms are interchangeable.

Filters work by blocking wavelengths of light associated with light pollution while passing light at wavelengths associated with objects the astronomer wants to either see or photograph. They work particularly well for objects that generate their own light at specific wavelengths, such as emission nebula. However, for objects such as galaxies, globular clusters, and reflection nebula they are less effective. The light generated or reflected by these objects is mostly broadband; thus, the light we want to filter out is at the same wavelengths as the light we want to keep (see Fig. 4.4).

Three basic kinds of light pollution reduction filters are available for the video astronomer to use:

- A broadband nebula filter. This filter passes blue and green light through an 80- to 90-nm wide band centered at approximately 485 nm and red light through a second band about 40-nm wide centered at approximately 660 nm. It may or may not pass infrared.
- A narrowband nebula filter, also called an ultra-high contrast filter (UHC), is similar to a broadband nebula filter except the lower bandwidth in blue/green is around 25 nm from 480 nm to 505 nm and red light through a second band about 40-nm wide centered at approximately 660 nm.
- A nebula line filter is the most precise, designed to pass narrowband emissions having bandwidths of 12 nm.

Line filters pass one very narrow band of the light spectrum, specifically at wavelengths of excited hydrogen and oxygen atoms. Three filters are popular:

- Hydrogen alpha with a 12-nm bandwidth centered at 656 nm (one version has a 6-nm bandwidth).
- Oxygen III (496 and 501 nm) with a 12-nm bandwidth centered at 501 nm.
- Hydrogen beta with a 12-nm bandwidth centered at 486 nm.

Perhaps the most useful nebula filter for a small aperture video telescope, say one having an aperture of 102 mm or less, is a broadband light pollution reduction filter. For larger apertures, a narrowband filter, also known as an ultra-high contrast (UHC) filter, is most often the better choice. Both filter types remove light associated with Mercury vapor as well as high and low pressure sodium street lights and natural sky glow. Both filter types pass light emitted by emission nebulae. Light transmission in the pass bands is around 95%, and the light transmission outside the pass bands is around 0% (see Table 4.3).

The difference between the two filter types is basically the bandwidth of light passed by the narrowband filter is not as wide as the broadband filter; thus, the narrowband filter removes more unwanted light and provides better contrast. The downside is that it also darkens the image more than a broadband filter, which is often problematic with telescopes having apertures of 102 mm or less. On the other hand, the broader pass band of a broadband filter attenuates less light than a narrowband filter providing a brighter image. The broader pass band also

Table 4.3 Broadband and narrowband light pollution reduction filter

Artificial sky glow			Emission nebula		Broad band nebula filter	Narrow band nebula filter
Mercury vapor	Sodium	Natural sky glow	Emission Lines	Wavelength	Pass band wavelengths	Pass band wavelengths
435.8 nm	589.0 nm	557.7 nm	H-β	486.1 nm	450–540 nm	480–525 nm
546.1 nm	589.6 nm		OIII	495.9 nm		
577.0 nm	615.4 nm		OIII	500.7 nm		
578.1 nm	616.1 nm		H-α	656.3 nm	650+ nm	645+ nm

allows more light from reflection nebulae and galaxies than does a narrowband filter, which is helpful when viewing or imaging these objects with an astro-video camera.

The choice between the two filter types for users of small aperture telescopes is often a difficult one to make. Regardless of which one is used, quality is very important. Some excellent broadband filters suitable for a small aperture telescope are made by companies such as Astronomik, Baader, Lumicom, and Thousand Oaks. There is one broadband light pollution reduction filter made of Neodymium that enhances viewing planets and the Moon but not nebulae and has little impact upon reducing artificial sky glow.

Earlier we discussed how an astro-video camera images objects having a surface brightness less than the sky glow in the middle of a large city with its ever present light pollution. Now take a look at how a broadband light pollution reduction filter improves the performance of an astro-video camera in the middle of a large urban area.

Recall that in the example nebula we explored the magnitude of sky glow was 17 per sq. arc-sec with a photon flux of 120 photons per second per sq. arc-sec. The peak brightness of the nebula had a magnitude of 18 per sq. arc-sec, with a photon flux of 48 photon per second per sq. arc-sec; the average brightness had a magnitude of 19 per sq. arc-sec, with a photon flux of 20 photons per second per sq. arc-sec; and the dimmest parts of the nebula had a magnitude of 20 per sq. arc-sec with a photon flux of 8 photons per second per sq. arc-sec.

Let's take a look at how a broadband light pollution reduction filter will improve the image of the hypothetical nebula in the example. Two exposure times are used for this example, 20 seconds and 40 seconds. The filter will pass light having frequencies within the pass bands and will not pass light for frequencies outside the pass bands. The pass bands of a broadband light pollution reduction filter cover approximately 50% of the visible light spectrum. So for this example, we assume the visible photon flux for sky glow will be reduced by 50% to 60 photons per second per sq. arc-sec. The transmission efficiency of the filter will be assumed to be 100% in the pass bands; thus, the photon flux from the nebula will remain unchanged. The results of this comparison are presented in Table 4.4.

The analysis shows that the use of a broadband light pollution reduction filter increases the signal-to-noise ratio of the image and increases contrast. When the exposure time is increased, contrast remains constant, but the signal-to-noise ratio increases. A similar analysis using a narrowband light pollution reduction filter will produce similar results but will have better contrast as well as higher signal-to-noise ratios than produced by a broadband filter.

This analysis comparing an image made with and without a broadband light pollution filter made some gross assumptions regarding the exact amount of light that is passed or attenuated as well as photon flux. The results are indicative of how either a broadband or a narrowband light pollution reduction filter improves the image made with an astro-video camera. However, the estimate of the magnitude of the impacts has room for error. What can be said is that a light

Table 4.4 Comparison between 20- and 40-second exposures with a broadband light pollution reduction filter

Exposure time	20 seconds	20 seconds	40 seconds
Photons per pixel	No filter	Broadband LPRF	Broadband LPRF
Sky glow	2400	1200	2400
Image, minimum	160	160	320
Image, average	400	400	800
Image, maximum	960	960	1920
Photon noise (min)	51	37	52
Photon noise (avg)	53	40	57
Photon noise (max)	58	46	66
Minimum image + sky glow	2560	1360	2720
Average image + sky glow	2800	1600	3200
Peak image + sky glow	3360	2160	4320
Effective signal/noise ratio (min)	3.1	4.3	6.2
Effective signal/noise ratio (avg)	7.5	10.0	14.0
Effective signal/noise ratio (peak)	16.6	20.9	29.1
Contrast (min)	7%	13.3%	13.3%
Contrast (avg)	17%	33.3%	33.3%
Contrast (peak)	40%	80.0%	80.0%

pollution reduction filter will increase both the contrast of an image and its signal-to-noise ratio.

Recall the three line filters briefly discussed, hydrogen alpha, hydrogen beta, and oxygen III. Each of these three filters has a very narrow pass band of only 12 nm, approximately 4% of the visible light spectrum. These filters produce very dark images and are not suitable for visual work in a small aperture telescope. However, for photography they are an excellent tool to combat light pollution. To properly use one of these filters a quality equatorial mount is needed that is capable of guided long exposures of around 20 minutes or more. The highest quality images are obtained with the astro-video camera set to the monochrome mode (night mode).

People tend to do what they know and often think of an astro-video camera in much the same manner as they would a DSLR or an astronomical CCD camera. This is a mistake. The reason is that exposure times with an astro-video camera are limited to very short exposures, 1/60th of a second (NTSC) or 1/50 of a second (PAL). By using the camera's integration and 3D-DNR features exposures equivalent to 85 (NTSC) and 100 seconds (PAL) can be obtained. This means that the signal part of an image made with an astro-video camera is low, and there is no way to increase the signal level by increasing exposure time.

What has this to do with filters and light pollution and astro-video cameras? Typically in astrophotography, the use of filters is accompanied by longer exposure times to increase signal levels. However, this option is not available for users of astro-video cameras.

The biggest impact is with line filters. With line filters, photographers typically make guided long exposures, sometimes as long as an hour or more. With long exposures, images made in urban areas using line filters can approach images made in very dark locations. However, an astro-video camera with a line filter will at best have a very low signal-to-noise ratio. Stacking literally hundreds of images is required to produce a signal-to-noise ratio adequate for a presentable photograph. This certainly cannot be used for near real-time visual work.

The inability of an astro-video camera to make long exposures also impacts the usage of broadband and narrowband filters much the same way as it does line filters but to a far less extent. The filtering of broadband and narrowband filters is much less severe and the signal-to-noise ratio is higher. Typically, a broadband filter is used for small aperture telescopes of 102 mm or less and narrowband filters for larger telescopes. Although the image is dimmed, adjustments in the camera's integration selection, brightness, etc., is usually possible for both visual and photographic applications.

Urban Viewing in Areas with Significant Light Pollution

As we said earlier, all telescopes love a dark sky. Unfortunately, most astronomers live in urban areas where dark skies only exist during wide area power outages. GOTO telescopes are almost a necessity in urban areas with significant levels of light pollution. The ever-present light dome drowns out all but the brightest stars to our unaided eyes, making manually finding objects difficult at best and impossible for many amateurs. However, the objects in the night sky are still there, and many are seeable if you can find them in your telescope. The trick is finding them. For this task, a GOTO telescope is essentially mandatory.

Once you get your telescope aligned, the mount will take you to any object you wish to see. Open star clusters and to a lesser extent globular clusters are excellent objects for viewing in urban skies. Splitting double stars is a good urban activity. Galaxies and nebulae are difficult. A wideband nebula filter such as the Astronomik CLS will help improve contrast, which helps you to see more details.

However, before you can find any objects in the sky with your GOTO telescope, you must align it. This often is a problem in an urban area, as the sky glow has drowned out all but four or five stars, and identifying the stars is often difficult. For alignment purposes, star charts often have too much detail. Many people find that a star chart showing only stars brighter than magnitude 1.5 is most helpful to identify the few stars that are visible in the night sky. Appendix C in this book provides urban star charts for the Northern Hemisphere at latitude 40 degrees north and the Southern Hemisphere at 30 degrees south, respectively. They should be very helpful in identifying alignment stars in an urban environment.

Chapter 5

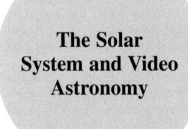

The Solar System and Video Astronomy

Our Solar System in Brief

The Solar System, our home in the universe. Is it unique, average, or very special? We really don't know. We only have one star, our Sun. In contrast, our nearest neighbor in deep space, some 4.37 light years away, is the Alpha Centauri system, a three-star system with one double star and another that may or may not be just passing by. One of the double stars is a G2V star, slightly bigger than our Sun. The other is a K1B star slightly smaller than our Sun. The third star, and the star closest to us, is an M5V red dwarf.

Let's assume our nearest neighbor has a planet in the habitable zone whose inhabitants are a bit more technically advanced than we are and who decide to visit our Solar System. How? That doesn't really matter. When our neighbors are about one light year from our Sun, they encounter the Oort Cloud, the most outer reaching part of our Solar System. The Oort Cloud is a sphere of icy objects surrounding the Sun with an outer boundary at about 50,000 Astronomical Units (AU). This sphere is some 30,000 AU thick. It contains several trillion objects less than 100 km (60 miles) in diameter that are composed mainly of water, methane, and ammonia. The total mass of objects in the cloud is estimated at about 10–100 Earth masses. The Oort Cloud, also known as the Opik Cloud, is the source of long-period comets. The existence of the Oort Cloud is based upon calculating the origin of the orbits of long-period comets. No objects coming from the Oort Cloud have been detected by telescopes on Earth, space telescopes, or space probes; thus, the Oort Cloud's existence and geometry is purely theoretical.

As our neighbors in space get closer, they encounter the Kuiper Belt, some 30–50 AU from the Sun. Like the Oort Cloud, the Kuiper Belt is composed of icy objects made of ammonia, methane, and water. It is also the home of periodic comets. More than 100,000 Kuiper Belt Objects (KBOs) of 100 km (60 miles) in diameter or less are thought to exist with over 1000 having been visually observed. The Kuiper Belt has three dwarf planets: Pluto, Haumea, and Makemake. The orbit of the dwarf planet Eris and its moon Dysnomia is three times further out than Pluto and is not part of the Kuiper Belt. Eris is the most distant Solar System object observed so far. The very largest KBOs are viewable with amateur telescopes.

Next on the journey from Alpha Centauri to our Sun are the planets Neptune, Uranus, Saturn, and Jupiter. These planets are gas giants, and along with their moons are of interest to many amateur astronomers. Inward from the gas giants is the Asteroid Belt, followed by the rocky planets: Mars, Earth with its large Moon, Venus, and Mercury. Only Earth falls within the habitable zone of the Sun. These planets and their moons as well as the Asteroid Belt are also of interest to amateur astronomers.

Finally the trip of 4.37 light years from the Alpha Centauri system to the Sun and its Solar System is completed. The Sun is a yellow dwarf on the Main Sequence with a spectral class of G2V. It is also a Generation I star; thus, it is rich in metals. Unlike most stars that are more massive than low-mass M stars (red dwarfs), our Sun is not part of a double- or multi-star system. Being the star that is closest to Earth, the Sun is the easiest star to study and is of interest to amateur astronomers (see Fig 5.1).

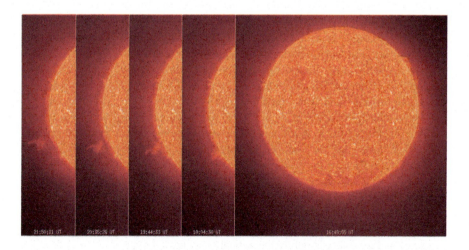

Fig. 5.1 Solar flare (Image courtesy of NASA)

The Advantages of Video Telescopes for Viewing the Sun

We have all seen the following warning or something similar about viewing the Sun:

VIEWING THE SUN IS THE ONLY ACTIVITY IN ASTRONOMY WHERE YOU CAN PERMANENTLY HARM YOUR BODY, SPECIFICALLY YOUR EYES. PERMANENT AND IRREVERSIBLE DAMAGE OR EVEN THE COMPLETE LOSS OF VISION, 100 % BLINDNESS, CAN RESULT.

The same warning also applies to your astro-video camera. The Sun can damage or destroy it just as quick and easily as it can damage your eyes.

Viewing the Sun is not an activity that a novice to astronomy should do. An Internet search using the terms "how to safely view the Sun in a telescope" will bring up several Internet sites that give specific information for safely viewing the Sun. Some things to consider are:

- Never view the Sun without using the proper full aperture solar filter or a solar telescope (see Fig. 5.2).
- Unless you are knowledgeable about solar viewing, never use a Herschel wedge, as the heat buildup can damage some telescopes.
- Never use dark colored-glass eyepiece filters, as they can crack from the Sun's heat, exposing your eye to the unfiltered light from the Sun, and the heat buildup can damage your telescope (see Fig. 5.3).
- Never use your telescope to project an image of the Sun, as heat buildup can damage your telescope.
- Always cap finder telescopes.

Fig. 5.2 White light full aperture glass filter

Fig. 5.3 Dark glass eyepiece filter

The Sun is very dynamic. It is rotating, but the rotation is not uniform—25 days at the equator and 36 days at the poles. The Sun is also in a continuous state of turmoil, with many constantly changing visible details. The most common feature visible on the Sun is its surface granulation, convective columns of hot plasma called Benard cells. These convective cells are often interrupted by concentrations of magnetic fields that cause dark, cool areas known as sunspots. Hot hoop or arc-shaped prominences, or filaments as they are sometimes called, erupt from the Sun's surface following the lines of magnetic fields. The Sun often burps, ejecting hot plasma into space. These burps are known as coronal mass ejections. Then there are magnetic storms known as solar flares on the Sun's surface that appear as very bright spots with a gaseous eruption. With all this, the Sun is enveloped by a sphere of plasma called a corona that is very visible during a solar eclipse.

Viewing or photographing the Sun, to say the obvious, is a daytime activity. During the day, the Sun heats Earth, providing the energy that drives Earth's weather. In this process Earth's surface is heated. This, in turn, heats the air near the surface of Earth, and the heated air rises. Refraction of light through the rising air is what produces the shimmering effect, or reflections such as standing water off in the distance, especially on hot summer days. This results in poor seeing conditions for viewing anything with a telescope.

Two options are available for safely viewing the Sun. The best option, if you can afford it, is a solar telescope specifically designed for viewing the Sun, such as one of Meade's Coronado series or one of Lunt's Solar System's telescopes. These telescopes have built-in hydrogen-alpha filters capable of safely revealing many details of the Sun. A less expensive solar telescope, the iOptron Solar 60, with a full spectrum filter, is also available.

Another option is to put a filter on an existing telescope. This is done with a full aperture solar glass filter specifically designed to fit the telescope or with a full aperture film filter. If you are handy you can easily buy solar safety film and build your own aperture filter. These full aperture solar filters as well as the one used on the iOptron Solar 60 telescope are known as white light filters and pass the entire spectrum of visible light from the Sun. Only about 0.001% of the Sun's light is actually passed through the filter; the rest is reflected, giving birth to the silvered surface of these filters.

White light filters are appropriate for viewing or photographing sunspots, surface granulation, solar eclipses, and transits across the Sun. However, these phenomena are only a small part of what is happening on the Sun. Solar flares, prominences, the solar corona, and other phenomena taking place in the Sun's chromosphere are invisible with a white light filter. To see this hydrogen-alpha (H-Alpha), calcium II K, or calcium II H filters are required. Hydrogen-alpha activity is visible to the eye and the camera. However, calcium activity is largely in the ultraviolet area at the lower limit of human vision and is not visible to many people. Also, ultraviolet light is harmful to the human eye. For these reasons, calcium filters are predominately a filter used for photography only.

The hydrogen-alpha filter is centered at 6562.8 Angstroms and has a very narrow bandwidth, typically one Angstrom or less. This is necessary to make chromospheric activity visible. The hydrogen-alpha filter does not need a white light filter to reduce the intensity of the Sun's rays that enter the telescope.

Note Hydrogen-alpha filters commonly used for photographing nebulae have a much wider bandwidth and pass too much light for solar work.

Calcium II K and H filters are centered at 3933 and 3969 angstroms, respectively. A broadband calcium II filter covering both the K and H bands is also available centered at 3950 angstroms, with a bandwidth of 80 angstroms. The broadband filter provides sufficient contrast for photographing granulation, flares, sunspots and other features prominent in calcium K. A white light filter is required when using a calcium filter.

A hydrogen-alpha filter is expensive and, dependent upon size and fittings, costs well over $1000 and upwards. The least expensive way to get a hydrogen-alpha filter is to purchase a solar telescope that comes with one. The Meade Coronado Personal Solar Telescope, a 40-mm f/10 refractor with a built-in hydrogen-alpha filter having a 1.00 angstrom bandwidth, is priced at $700. Lunt Solar Systems offers a 50-mm f/7 dedicated hydrogen-alpha refractor with a 0.75 angstrom bandwidth starting at $800. The lowest price calcium K filter starts at $360 for the 1.25-inch Baader broadband filter. However, narrowband calcium II K filters are just as pricy as hydrogen-alpha filters.

Unlike deep space, the Sun is a dynamic object to photograph. It rotates, and its surface features are in a continuous state of seemingly random change. Not only do features change continuously, the rate of change is often rapid. This is significant from a photography perspective, as the rate of change is frequently sufficient enough to blur images when exposure times approach one minute. If all this is not enough, solar imaging is done during the day, when seeing conditions are at their worst.

In spite of the fact that the Sun is extremely bright, seeing conditions on Earth greatly impact the view of the Sun, either by an eyepiece or by a camera. The churning upwelling of heated air during the day means that the best viewing time is near solar noon, when the Sun is high in the sky and the amount of atmosphere penetrated is at a minimum. Another good time for viewing is early in the morning, before the Sun has a chance to heat the air. The problem with early morning viewing is that the Sun is low in the sky; thus, its image is distorted by the amount of atmosphere its light rays must pass through. Viewing in the afternoon is generally the worst time of the day. Convective currents of hot rising air distort images. Add to this the long path of Earth's atmosphere that must be penetrated. These two problems combine, and the result is an extremely poor environment for viewing or for photography.

You will need to decide for yourself which time you like best. Either way, early morning or noon, Earth's atmosphere has two major impacts upon viewing and photographing the Sun. Obtaining an exact focus is difficult to do and, even when achieved, the churning atmosphere distorts the image.

From a photography perspective, photographing the Sun is not difficult, but making an outstanding image is. This is because:

- Observations are in the daytime, which typically has bad seeing conditions due to the heat from the Sun.
- Focusing is difficult due to the bad seeing conditions.
- Some features, particularly those visible in an H-alpha filter, can change sufficiently in a short period to produce "soft" images after a minute of elapsed exposure time.

These conditions, while not ideal for any camera, provide an environment that plays into the strengths of an astro-video camera. Even though the real-time display of an astro-video camera facilitates focusing, the bad seeing conditions often makes knowing when focus is achieved difficult, as the Sun fluctuates in and out of focus.

For photography, an astro-video camera offers a way to work around the focus issue as well as the bad seeing conditions. It produces 1800 frames per minute (NTSC) or 1500 frames (PAL). Do a 30-second imaging run. Sort these frames, and only stack the ones with the sharpest images. This produces a final image with sharp details if the camera is focused. A variation of this technique is to make three series of images. First make a series of images with the camera focused as best you can achieve, and then repeat with the telescope defocused a tiny amount either side of the focus point. For the final image, use the series that has the best images.

However, not everyone wants to photograph the Sun. Some enjoy viewing the Sun and others prefer sketching to photography. Although an astro-video camera is well suited for capturing images of the Sun, TV monitors and computer screens aren't particularly good for displaying anything outside in broad daylight. The light from the Sun easily washes out their displays, and the astronomer can see very little (Fig. 5.4). Several workarounds are used, such as:

- a simple fabric scrim such as a blanket thrown over the observer's head and TV monitor/computer screen, equipment.
- equipment tents that cover only the TV monitor or computer screen.
- a small tent outside near the telescope to house the equipment and observer.
- locating the TV monitor inside your house, garden shed, or garage, for example.

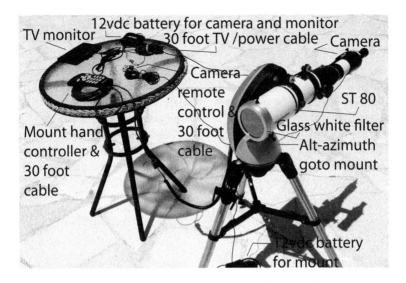

Fig. 5.4 White light video solar telescope shows a typical white light video solar telescope. This particular kit has a 30-foot (9-m) long camera remote control cable, a 30-foot long telescope mount hand controller cable, and a 30-foot-long RG59 coaxial cable. This allows the display to be located in a nearby room, garage, shed, etc., which greatly improves the ability to observe images on a TV monitor in the middle of the day

Recall from Chap. 1 that unlike the digital signals from web cameras and astronomical CCD cameras that are limited to about 4.5 m (15 feet), the analog signal from an astro-video camera can easily travel upwards of 30 m (100 feet) and much more if quality cables are used. This provides a lot of flexibility regarding the location of the display equipment.

Locating viewing equipment inside a house has several advantages in addition to the comfortable viewing environment created by getting away from the heat and bugs of summer and the cold of winter. The large screen display of the family's television has its advantages. Several people can see the view of the Sun at the same time, though at the cost of image sharpness. If line losses due to the length of the coaxial cable are not too bad, the output from the camera can also be split with one feed to a TV monitor and the other to a computer via a frame grabber. Once the signal is in the computer, the image can go anywhere in the world where Internet service is available, making the images available real-time to science classes, public outreach programs, distant relatives, etc.

One interesting characteristic of both the Sun and the Moon is their apparent size in the sky is essentially the same, about 30 arc minutes. As discussed in Chap. 2, the physical sizes of sensor chips typically used in video cameras are very small; especially in relation to the focal length of telescopes. Small sensors and long focal lengths produce images with small fields of view (see to Table 2.4). This characteristic also produces large images; thus, the focal length of the telescope used is an important factor, especially if you want to make a full image of the Sun or the Moon.

Often the field of view provided by a particular camera/telescope combination is too small to capture the full image of the Sun or the Moon. Thus, a mosaic is used if a full image is desired. Seamless solar mosaics are difficult to make, as the Sun's dynamic surface and the rapidly changing seeing conditions of the day impede such efforts. The Moon is less impacted by seeing conditions, and its physical features are permanent. However, like the Sun, it is rotating, and shadows are constantly changing. These attributes preclude spreading the imaging of either the Sun or the Moon over multiple sessions when making a mosaic.

Video Telescope Attributes for Exploring the Moon

Lunar Observation

The Moon, our nearest neighbor in space, is one of the first objects people see in a telescope, vying for attention with Saturn and Jupiter. The Moon is very bright and not impacted by today's ever present light pollution. It is available anytime it is visible in the night sky and a prime object for viewing for some people located in urban areas. Unlike Saturn and Jupiter, even a very modest telescope or binoculars show sharp details of the lunar surface (see Fig. 5.5).

Fig. 5.5 The Moon (Image courtesy of NASA)

There are a lot of activities you can do related to observing the Moon. Some like to sketch prominent surface features that they see in their eyepieces. Then there is a sizable group who simply like to relax, find, and then study specific permanent lunar features such as craters, maria, mountains, valleys, rims, and so on. Also of interest are transient optical features. Two popular ones, known as lunar X and lunar V, are only visible under specific lighting conditions for a few hours near the first quarter of the Moon.

Although there are no reasons why a video telescope cannot perform very well for these viewing tasks, many astronomers, perhaps most, prefer the view through an eyepiece rather than on a TV screen. However there are lunar activities where a video telescope is a most valuable tool.

Lunar Photography

Lunar photography is one area where an astro-video camera functions remarkably well. The photographer can literally make thousands of images with an astro-video camera. Each image is then examined, and the very best are selected and combined into the final image. This method produces a sharp and detailed image not attainable by typical film or digital cameras, whether it is some particular geological feature or area of the Moon or even a mosaic of the Moon's surface that is visible from Earth.

Imaging the Moon's surface is in some ways similar to that of imaging the Sun. Although not nearly as severe as for solar photography, atmospheric seeing distorts the camera's view of the Moon. Obtaining a precise focus for the Moon is not as difficult as it is for the Sun but does take some effort. Since the Moon focuses at infinity, a *Bahtinov mask* on a nearby star is used by many. An astro-video camera produces images at a rate of 25–30 frames per second (PAL or NTSC); thus, the image on the computer screen provides real-time feedback for focusing. Regardless of how you focus, spend some time and obtain a sharp focus.

How many frames should you capture? An astro-video camera produces 1800 frames per minute (NTSC) or 1500 frames (PAL). These frames require a sizable amount of memory for storage. However, there are other factors to consider. One is the mount. An alt-azimuth mount is subject to field rotation. Depending on the location of the Moon in the sky and the location of the telescope on Earth, field rotation may limit the total time available for exposures as short as 20 seconds before field rotation impacts the image. The 20 seconds produces 500–600 frames (PAL or NTSC). If your stacking program can adjust for field rotation, then you can work around this issue and obtain as many exposures as you wish. A precisely aligned equatorial mount will not have this issue, and exposure times of several minutes are permissible. If you are using an alt-azimuth mount, Appendix B in this book has tables that provide total allowable exposure times based upon the locations of the object being photographed and the location of the camera.

Lunar Impact Monitoring Program

One area where an amateur can make valuable contributions to science is by participating in NASA's Lunar Impact Monitoring Program. The objective of this program is to determine the rates and sizes of large meteoroids a few centimeters in size striking the surface of the Moon. The program is managed by the Meteoroid Environment Office at the Marshall Space Flight Center in Huntsville, Alabama, in the United States. Details of the program are available at: http://www.nasa.gov/centers/marshall/news/lunar/program_overview.html#link1.

To participate in the program you will need a video telescope having the following specifications:

- a 200-mm (8 inch) telescope with approximately a 1000-mm (40-inch) focal length
- an equatorial mount or derotator tracking at lunar rate
- an astro-video camera with a ½-inch sensor
- a digitizer for digitizing video and creating a 720x480 .avi compatible with LunarScan (see below)
- a time encoder/signal
- a GPS timestamp or WWV audio
- a PC compatible computer with approximately 500 GB of free disk space
- software, such as Windows XP or higher; WinDV, used for recording .avi, vid+auds setting; LunarScan, used to locate impact flash candidates in an .avi; VirtualDub, used for making flash and stellar calibration video clips; Limovie, used for video photometry; and a virtual Moon atlas, used to locate the position of the flash on the Moon.

The LunarScan software is a free download from the Meteoroid Environmental Office. It is available only on the Windows platform, and no plans exist to port it over to other operating systems.

Except for unusually large meteors, webcams and camcorders do not have the sensitivity required to detect meteor strikes on the Moon. A frame rate of 25–30 frames per second is required to capture most strikes, as the duration of the flash of light from the strike is often less than 1/30 of a second. The total exposure time for each frame is equal to the frame rate. This precludes any dead time between frames and also prohibits integration with a CCD camera.

Observation is restricted to certain times of each lunar month. For example, observations are typically made when the Moon is less than 50% illuminated. However observations are not made during the waxing and waning phases of the Moon. During these phases the Moon is too close to the Sun and low in the sky, where the Sun's glare and atmospheric extinction reduce the ability to detect the faint flashes of light caused by a meteor strike.

Reporting a suspected meteor strike is relatively simple to do. An email to the Meteoroid Environment Office is all that is required. The email must include the date, time, observatory location on Earth, and the impact location on the Moon. Two files are required as attachments to the email. One is a file containing a few frames of video either side of the flash, with the time of the flash designated. The time must be accurate to within approximately 0.1 second of Universal Time. The second file is a few frames of a star near the Moon made within a few "tens of minutes" of the flash along with the star's identification. The Meteoroid Environment Office uses the star to photometrically calibrate the flash.

Occultations

A third lunar activity very suited for video astronomy is timing occultations of stars by the Moon. An occultation is when some object passes between a star and an observer that blocks the view of the star. Two types of lunar occultations exist:

1. A total occultation is when the observer sees only one event, either a star disappearing behind the Moon as it passes in front of the star or the star reappearing as the Moon moves away from the star.
2. A grazing occultation is when the Moon just grazes a star and the star blinks in and out of view as it passes behind lunar mountain peaks.

Of the two types of occultations, a total occultation is by far the easiest to observe and measure.

Light from a star is parallel; thus, the Moon's shadow on Earth from the star covers an area equal to a cross sectional area of the Moon. An occultation can last several hours, so the path on Earth where a total occultation event is visible covers a very large area. The same is not true for a grazing occultation. A grazing occultation only occurs at the Moon's north and south poles and is visible from Earth along a path a few kilometers wide (a couple of miles) at the northern and southern borders of the total occultation path. Although a fixed observatory can record many total occultations during a year, it will be in the path of a glazing occultation only a few times at best each year. For grazing occultations, the observer typically must travel to the geographic location on Earth where the grazing occultation is visible.

Although an occultation can be observed anytime during the lunar month, the glare from the reflection of the Sun can make detection of the event difficult and introduce possible timing errors. The leading edge of the Moon is not illuminated by the Sun during the first two weeks of the lunar month. This is the best time to observe a star disappearing behind the Moon, as the Sun's glare will not interfere. On the other hand, the trailing edge of the Moon is dark for the last two weeks of the lunar month, making it the best time to observe a star's emergence from behind the Moon.

Observing and measuring occultations, like the lunar impact monitoring program, is an area where amateur astronomers can make meaningful contributions to science. If occultations interest you, the International Occultation and Timing Association, also known as IOTA (www.occultations.org), is an excellent resource for information on when and where occultations are visible, how to observe an occultation, the equipment needed, and the process for reporting results. The IOTA recommends that beginners start with observing total occultations.

To observe an occultation, a small telescope with an aperture of 80 mm (3 inches) or more will suffice. For a visual observation, an accurate timing source for universal time is required as well as a stopwatch and a way to record the event by audio. An accuracy of ±0.1 seconds is needed.

Astro-video cameras are popular for observing occultations because they can produce timing accuracies not attainable with manual methods and also because they can operate as remote observing stations that are easy to configure and operate. The number of possible configurations using an astro-video camera is quite large. One simple but very capable configuration contains the following major components:

- astro-video camera
- small aperture telescope
- GPS unit
- GPS video time inserter
- small camcorder with an LCD monitor screen
- portable power supply

This configuration is excellent not only for total and glazing occultations by the Moon but also for observing occultations by planets and asteroids.

Fig. 5.6 Jupiter (Image courtesy of NASA)

Video Telescope Attributes for Viewing the Planets

"I want to see the Moon and the planets." That is a statement frequently made by people just beginning to explore the night sky. The Moon, Saturn, and Jupiter seldom fail to live up to expectations (Fig. 5.6). However, after a few quick looks most head into deep space, leaving our neighbors behind—seen that, done that, so to speak.

Planetary observing is a dynamic activity. Many astronomers study the motion of planets in the night sky. The surface features of three planets—Mars, Jupiter, and Saturn—are easily observable from Earth and also offer a continuing change in surface details. Perhaps the most noted is Jupiter, with its turbulent and ever changing patterns and color values in its cloud bands. Its famous Great Red Spot varies in shade and sometimes disappears altogether for a short while. Then we have the continuous dance of its four main moons as they orbit, casting shadows on Jupiter's surface or disappearing completely as they pass behind the giant planet.

Saturn puts on a similar but less dynamic show, as it, too, has clouds bands and clouds to observe. However, its main show is its rings, which are the subject of study by many amateurs. About every 26 months Mars passes close by making its surface features and polar cap visible even in small aperture telescopes of around 4 inches or more. For a few weeks, it is in the limelight as amateurs study its clouds and try to ascertain its surface features.

Venus and Mercury are very bright and visible in both the morning and evening sky. Both planets show as crescents, like our Moon; but, unlike the Moon, surface details are blurry at best. In the year 2014 Uranus surprised both amateur and professional astronomers as bright spots became visible on its blue surface. Professional astronomers used photographs made by amateurs to help justify a target of opportunity for the Hubble Space Telescope, which then imaged the entire planet. Neptune and the minor planets, such as Pluto, are but tiny dots in a telescope. Their primary interest for amateurs is the measurement of their movements.

Astro-video cameras play two roles regarding planets. The most well known and widely used role is making images of the planets. The same attributes that make these cameras suited for imaging the Sun and the Moon are also useful for producing sharp images with great detail of the planets. As with lunar and solar photography, a few thousand images are made, and the best stacked to provide a final image. While photographing planets, the photographer must keep in mind that these planets are rotating as much as 36 degrees per hour. Care must be exercised to not stretch the duration of the imaging session too much.

One thing to note about planetary imaging is that the size of planets as seen from Earth is very small. Recall the discussion in Chap. 2 concerning the relationship between astro-video camera sensor size and telescope focal length to image size and field of view. Planets have a small apparent size viewed from Earth, as shown by Table 5.1.

Table 5.1 Apparent size of solar system objects

Solar system object	Angular diameter	
	Min	Max
Sun	31.6'	32.7'
Moon	29.3'	34.1'
Mercury	4.5"	13.0"
Venus	9.6"	66.0"
Mars	3.5"	25.1"
Ceres	0.3"	0.8"
Vesta	0.2"	0.6"
Jupiter	29.8"	50.1"
Saturn	15.0"	20.8"
Uranus	3.3"	4.1"
Neptune	2.2"	2.4"
Pluto	0.06"	0.1"

For photographing planets with an astro-video camera, an eyepiece projection or a Barlow lens is often used to increase the effective focal length of a telescope, thus producing a larger image of the planet.

There is a lesser appreciated benefit that an astro-video camera brings to planetary observations. Many features on planets are too faint or the color values too similar for detection in an eyepiece. These features are easily detected by an astro-video camera, providing information to astronomers that would not exist otherwise.

Astro-Video Camera Attributes for Photographing Smaller Solar System Objects

Scattered throughout the Solar System are small objects called asteroids, made mostly of rock and dust and a scattering of ice. Most asteroids orbit the Sun in belt known as the Asteroid Belt, located between the orbital paths of Mars and Jupiter. Nearly all asteroids are very small, ranging in size from small boulders to about a kilometer. However, some are considerably larger. The largest asteroid is a dwarf planet named Ceres, with a diameter of approximately 950 km (590 miles). Three others, Vesta, Pallas, and Hygiea, are 400 km (250 miles) or longer in size. These objects are visible at times in binoculars, and Vesta can be a naked eye object under a dark sky. Many asteroids cross Earth's orbital path and many strike Earth.

Asteroids, especially asteroids with near Earth orbits, are of interest to many amateur astronomers. Some astronomers participate in a program known as the Target Asteroid Project established by NASA. This outreach program supports the "Origins Spectral Interpretation Resource Identification Security—Regolith

Explorer (OSIRIS-REx) Mission" that includes visiting an asteroid and returning with a sample from its surface. The objectives of the OSIRIS-REx mission are to:

- make a direct measurement of the Yarkovsky effect (thrust provided by thermal radiation) on an asteroid named Bennu that has a significantly high probably for striking Earth in the year 2182.
- to determine Bennu's physical and chemical properties to support a future impact mitigation mission, if needed.
- to obtain pristine carbonaceous material from the early Solar System to study for clues related to the origin of life.
- to gain a better understanding of asteroid orbits through ground-based observations.

This last element involves amateur astronomers who select and observe asteroids from a list published by NASA. The Target Asteroid Project's imaging requirements—three, thirty-minute exposures—exceed the capability of astro-video cameras.

Another activity of interest to many amateurs is observing occultations. The process and equipment are essentially the same as used for observing lunar occultations. The major difference between the two is that the path an asteroid occultation makes across the face of Earth is much narrower. Several observation stations set up across the asteroid's path can even produce information useful for obtaining the overall shape of the asteroid. As with lunar measurements, the results of asteroid occultations are submitted to the International Occultation and Timing Association.

Chapter 6

Deep Space and Video Astronomy

Our Window into the Universe

Deep space is our window into the universe and where most amateur astronomers spend their time. It is a wondrous place. Everywhere we look, we see countless numbers of galaxies, each with countless numbers of stars. Vast nebulae made of gas and dust clouds span light years across space.

Nebulae give testimony to the end of dying stars as well as to tell a story of stellar nurseries giving birth to the next starry generation. Powerful energy jets shoot from the heart of many galaxies, including our own Milky Way. Some stars die in a violent burst of energy, while others slowly flare and fade away. Here in our part of the Milky Way we see that our star, the Sun, is a solitary star, yet our nearest neighbor in space, the *Alpha Centauri system,* is a triple-star system. All in all, deep space is full of places to go and things to see or photograph. In reality, it's the wonders of deep space that attract most amateur astronomers to astronomy.

One very popular use for video telescopes is simply enjoying viewing and exploring objects in deep space. This is an activity as old as telescopes and began when Galileo pointed his small refractor towards the heavens. Optical telescopes of today provide untold hours of enjoyment to people all over the globe. This aspect of astronomy recently took another step forward. Improvements in camera technology now allow the marriage of telescopes with television. Today we can substitute the telescope eyepiece with a video camera and see even deeper into space and see in color and comfort as well (see Fig. 6.1).

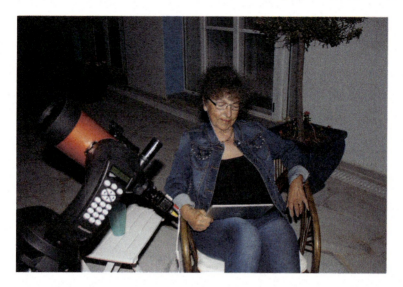

Fig. 6.1 Video astronomy with a 4SE

For viewing deep space optically, aperture rules. As aperture increases, fainter objects are seeable with higher resolution, if all else remains equal. Unfortunately financial, physical, logistics, and other constraints place limits on the size and kind of telescope a person can have. Only a few people can afford a telescope having an aperture of 12 inches or larger or a high quality 5- or 6-inch refractor. An 8-inch telescope is perhaps the most popular aperture for amateurs. It is often called the "Goldie Locks" telescope, affordable yet big enough to see details in deep space while not too large or too heavy to handle. A high percentage of astro-video astronomers use the same telescope for video astronomy that they use for visual work. The result is that 8- and 10-inch f/10 Schmidt Cassegrain telescopes are very popular and play a major role in video astronomy.

An astro-video camera can conservatively increase the effective aperture of a telescope by a factor of two to three times. This is a statement repeated often by advocates of video astronomy. What they are referring to is the apparent magnitude of an object being viewed in a telescope. For example, a 4-inch telescope with an astro-video camera can display objects having apparent magnitudes that normally require an 8- to 12-inch telescope to view using an eyepiece. The 8-inch telescope will have the ability to show objects that normally require a 16- to 24-inch aperture for viewing with an eyepiece. A lot of factors govern this aspect of video astronomy, such as the darkness of the sky, the quality of the mount, and the settings used for the camera, but a doubling of capability is not difficult to obtain.

However, the apparent magnitude of an object that a telescope can detect is only one attribute related to its aperture. The other aspect is the resolution of an object. Resolution is dependent upon a telescope's aperture and its optics. If you include a

Fig. 6.2 M3 viewed through an 80-mm f/5 refractor

camera, another variable is added—the camera's resolution. Although an astro-video camera can show fainter objects than are seeable with the eyepiece of a telescope, the resolution of the astro-video camera display is limited by the resolving power of the telescope as well as that of the camera. Since astro-video cameras typically comply with either the NTSC or PAL television standard, their resolution is rather low in comparison to digital cameras in general. The end result is the images produced by an astro-video camera often do not have the resolution of images seen in an optical telescope. However, the image produced is in color and will show features requiring an aperture two to three times greater than the one being used. Take a look at Fig. 6.2 M3 viewed through an 80 mm f/5 refractor. The image used an integration of X128 and the 3D-DNR features of the camera. This produced a time lag of approximately 11 seconds which is not real time but is close enough for visual observations.

A near real-time image of a deep space object produced by a video telescope will not look like an astrophotograph of the object typically produced by astronomical CCD or DSLR cameras. This sometimes disappoints first-time viewers using a video camera. This is not to say that an astro-video camera cannot produce images similar to those made by a CCD or a DSLR camera. Many users of astro-video cameras download images into a computer and then stack and enhance them using the same programs and techniques used by traditional astrophotographers. The results obtained, especially for the brighter DSOs, such as the Messier and Caldwell objects, are often very good and competitive with traditional methods. But for now, let's look at video astronomy only. Video astrophotography is an area of its own and covered in the next chapter.

Issues Related to Deep Space Objects and Video Astronomy

A video telescope will have the following four major components:

- video camera
- telescope
- telescope mount and tripod
- video display.

Each of these components must be integrated into one system to produce an output having acceptable characteristics. This means we must match the parameters of the individual components to produce a final image having the desired size, brightness, and resolution. This is vague but will become clear later on.

Some of the major attributes of an astro-video camera are its sensor size, low light sensitivity, and noise reduction features. The sensor size of most astro-video cameras is small. A typical entry/budget-level camera will have a 1/3 of an inch sensor, while the more expensive cameras may have a ½-inch sensor. As discussed in Chap. 2, these small sensor sizes have a major impact upon the field of view and image size produced by an astro-video camera.

The focal length of a telescope is an important attribute that impacts the field of view and image size produced by an astro-video camera. The longer the focal length, the smaller the field of view, and the larger the image size, other parameters being equal. Another attribute, the focal ratio of a telescope, impacts image brightness. The lower the focal ratio, the brighter the image will be and the shorter the exposure time that is needed or the greater the contrast. Exposure time can influence the telescope mount chosen, as shorter exposure times can lessen the importance of mount tracking and stability.

As with astrophotography with an astronomical CCD or a DSLR camera, the mount used for video astronomy is of major importance. The images produced are directly correlated to the stability and tracking ability of the mount. The preferred mount is a German equatorial mount on a pier in a permanent observatory. However, that is not possible for most people, who must use a tripod-supported mount and do not have an observatory or must travel to an observing site away from their homes. An alt-azimuth mount can be used but will have issues related to field rotation that must be considered. This is especially true with current cameras and their long integration periods. Theoretically, a manually operated, motorized, mount can be used; in reality, though, and for practicality a modern GOTO mount is best.

The analog video output of current astro-video cameras complies with either the NTSC or PAL standard and is fixed when the camera is manufactured. You cannot switch between the two standards. This means that your video display device must be capable of matching the standard used by the camera. Since the output of the video camera is either an NTSC or a PAL standard output, the video display device (TV set, TV monitor, DVD player, VCR recorder, etc.) must be compatible with it. Fortunately most modern display devices can sense which standard is used and match either one. Another impact caused by the NTSC or PAL standard is resolution.

The images are not high definition TV images and, even if converted into high definition format, will not have a high resolution.

If you are using a computer or tablet as a display device then you will need the video capture device called a frame grabber to convert the analog video signal into a digital signal for the computer. The frame grabber must be compatible with the analog video signal as well as the computer and its operating system.

Another aspect to consider is the size of the monitor used to view the images produced. A 9-inch TV monitor is popular for solitary viewing, big enough to show details and share with one or two people yet small enough to remain portable and have a reasonable battery life. A very small display device, say, a 4-inch TV monitor, will appear to have a sharp image but will show few details. At the other extreme a 50-inch TV will have a large image that can be viewed simultaneously by a large number of people. However, the details will appear soft, and finding a power source for the monitor is often problematic.

Camera-Related Issues

Currently three budget/beginner-level astro-video cameras are available on the U.S. market. These cameras, as well as some intermediate-level cameras, use a common sensor having the following physical characteristics:

- sensor size: 1/3-inch
- sensor measurements: 4.8×3.6 mm
- sensor diagonal: 6 mm
- total number of effective pixels: 480,000 (976 vertical $\times$ 494 horizontal)
- pixel measurement: 0.0050 mm vertical $\times$ 0.0074 mm horizontal
- analog video output.

The above characteristics are for a camera manufactured to the NTSC standard. Cameras manufactured to the PAL standard will differ somewhat with the total number of pixels and pixel size; however, the differences are not significant.

Recall from Chap. 2 that modern, entry-level, budget astro-video cameras are integrating cameras that have extensive noise reduction features referred to as 3D digital noise reduction (3D-DNR). These cameras use electronic shutters with shutter speeds ranging from as short as 1/100,000 of a second to as long as 1/60 of a second NTSC or 1/50 of a second PAL standard. Yet, they are capable of an integrated exposure time equivalent to a single exposure as long as 17 seconds NTSC or 20 seconds PAL. They do this by linearly stacking up to 1024 exposures. If the 3D-DNR feature is used, then there are up to five linearly stacked images. This increases the maximum effective exposure time to 85 seconds NTSC or 100 seconds PAL.

These times are sufficiently long enough that the impact of field rotation must be considered when capturing an image. One aspect of 3D-DNR used by current entry-level cameras is the ability to adjust for some translational and rotational movement. Actually the feature betrays the CCTV ancestry of astro-video cameras, as they were designed to enhance the image of moving objects such an automobile or a running person.

Another feature available is the Stabilization Option in the OSD EFFECTS menu, which is similar to image stabilization in today's DSLRs and binoculars. Image stabilization and 3D-DNR can mitigate small amounts of field rotation as well as tracking movement such as produced by a poorly aligned equatorial mount. However, these features have limited usefulness for significant amounts of field rotation typical of alt-azimuth mounts. Although an alt-azimuth mount can be used with good results, field rotation can produce softer images with slightly larger stars than the images produced by an equatorial mount. For visual work, the difference between an equatorial mount and an alt-azimuth mount is hardly noticeable.

The output of a budget, entry-level astro-video camera is an analog video output that complies with either the NTSC or PAL standard. This is a low resolution standard. If the image is exported to an external computer for processing, current video capture devices do an 8-bit conversion for the MS Windows platforms. Interestingly, internally the camera does a 10-bit analog to digital conversion for pixel data and 16 bits for internal processing. This implies that the more you use the camera's capabilities for data processing, the higher the quality of the final image obtained will be. Keep in mind that this is done using exposures that cannot exceed 1/60 or 1/50 of a second in duration, dependent upon if you are using a camera designed to the NTSC standard or the PAL standard.

The size of a camera sensor has a major impact upon imaging deep space with an astro-video camera whether or not it is used for visual work or for photography. A 1/3-inch sensor has a diagonal measurement of 6 mm. The focal lengths of telescopes typically used for astro-video are often very long. An 8-inch SCT has a focal length of 2000 mm. A small sensor size and a long focal length produce images that have a very small field of view and a large image size that is often dim. This creates several issues that relate to finding and tracking objects as well as viewing them. An image with a small field of view and large image size will only show a portion of an object. While there are times when this may be desirable, typically observers like to view an entire object at once.

Small movements of the telescope are amplified; thus, a solid mount with accurate tracking and no gear-induced movement is required if high quality images are desired. Finding objects is often difficult as the small field of view is akin to viewing through a soda straw. This makes star hopping very difficult and for most people requires an accurate GOTO mount. This is especially true for newcomers to astronomy.

Image resolution is a factor of the telescope and the camera. All else equal, the larger the aperture of a telescope the greater the resolution. The camera's resolution is dependent upon the sensor's physical size, which is governed mostly by the number and size of the pixels used by the sensor as well as how tightly the pixels are packed. Other factors, such as whether or not the camera is monochrome or color, etc., are also involved.

With the exception of some refractors, most telescopes used for visual observing today have a focal length of 800 mm or more, and many also have high focal ratios.

This means that the images produced by an astro-video camera with a typical telescope often will have a narrow field of view and a large but dim image. Neither of these attributes is desirable for viewing deep space objects where image brightness and field of view are important attributes in enhancing the ability to observe distant, dim objects often spanning light years in space.

The ubiquitous 8-inch f/10 Schmidt Cassegrain telescope (SCT) is widely used for video astronomy. This telescope is found on single and dual tine, alt-azimuth, fork mounts as well as German equatorial mounts. The SCT is a compound telescope (has mirrors and a lens) and, unlike most refractor telescopes, produces images essentially free of chromatic aberration. The basic optical characteristics of an 8-inch SCT are as follows:

- aperture: 203 mm (8 inches)
- focal length: 2030 mm (80 inches)
- focal ratio: f/10

An astro-video camera with a 1/3-inch sensor and an 8-inch f/10 telescope combination will have a field of view of approximately 8×6 arc minutes. This is a very small area of the sky and will allow only a partial view of all but the smallest deep space objects. Finding objects with such a small field of view is often a chore, especially objects not visible in an eyepiece. The combination will also be sensitive to mount movement and vibration. At f/10, the image will be dim. These issues are not conducive to producing an enjoyable observing session; on the contrary, they are often the source of much frustration, especially for beginners to video astronomy. Fortunately these issues are resolved using focal reducers to shorten the effective focal length of an 8-inch SCT to around 680 mm and to decrease the focal ratio to around f/3.3. Otherwise, video astronomy would not be gaining in popularity as it is now.

Viewing Deep Space with a Video Telescope

How do we view the universe with a video telescope? Actually, the process is not difficult once you get the hang of it. However the first few times out often have some frustrating moments, even for experienced astronomers and astrophotographers. For beginners to astronomy, spending some time to become familiar with the night sky and observing with a typical telescope will make the task much easier. The simple truth is that if you do not know what is in deep space and how to set up and use a telescope, the likelihood of finding and observing an object with an video telescope is very small.

Three budget astro-video cameras are currently available on the commercial market: the Astro Video Systems (AVS) DSO-1 camera, the MallinCam Micro EX camera, and the Revolution Imager. All three cameras are popular with beginners. These cameras are based upon the same technology, have very similar capabilities, and are competitively priced at around $200 including accessories. The operation of all three cameras is very similar.

The type of mount, alt-azimuth or equatorial, has little impact upon the process used for visually viewing deep space with an astro-video camera attached to a TV monitor. Generally thought more complex to use than an alt-azimuth mount, an equatorial mount does provide slightly sharper near real-time images of many deep space objects. Also total integration times are not subject to constraints caused by field rotation. However, an inexpensive alt-azimuth GOTO mount with a fast telescope provides very acceptable views of deep space objects. This is especially true if the viewer has no desire to make photographs of the deep space objects seen in the telescope. Quite often seeing conditions are more critical than the type of mount that is used for visual observations.

The possible telescope and mount configurations available for video astronomy are quite large. An excellent budget or starter kit is the Orion ST80A or the SkyWatcher StarTravel 80-mm f/5 short tube refractor on a German equatorial mount such as the iOptron SmartEQ PRO or the SkyWatcher EQ3PRO if astrophotography is also on the addenda. For viewing only, alt-azimuth mounts such as the Celestron SLT and Prodigy mounts, the SkyWatcher SynScan AZ GOTO mount, the Orion StarSeeker IV mount, and the iOptron Cube series mounts are viable alternatives to an equatorial mount. The GOTO features and capabilities of all these mounts are acceptable for video astronomy.

Both of the 80-mm f/5 short tube refractors mentioned have quality optics and a solid metal 1.25-inch focuser (see Appendix D in this book for ways to tune up the focuser). They are simple to use, and their very short focal length not only provides a large field of view but also makes it forgiving of mount movement and vibrations. At f/5 they have a bright image. A 0.5x focal reducer lowers the focal ratio to f/2.5, which produces a smaller but even brighter image. This also reduces the focal length from 400 mm to 200 mm. With a focal length of 200 mm, a 1/3 of an inch sensor can provide a field of view that is 83 × 62 arc-minutes in size. To put this in perspective, a field of view produced by an astro-video camera having a 1/3rd of an inch sensor with a short tube 80-mm f/5 refractor using a 0.5x focal reducer is comparable to the field of view produced by a DSLR with a APS-C sensor through an 8-inch Newtonian at f/5. The resultant field of view is sufficient to capture nearly all of a large object such as the Lagoon Nebula, M8, which is 90 × 40 arc-minutes in size.

Any video telescope kit consisting of the above components provides a good budget kit for starting video astronomy or for someone who just wants to be involved with video astronomy without spending a lot of money. The mounts have sufficient accuracy and the telescopes have the focal length and photographic speed needed to provide pleasing images in near real-time. The combination of any of the mounts or telescopes will be very forgiving regarding vibration, wind, and so on.

One aspect of video astronomy that sometimes perplexes newcomers is the importance of a telescope's focal length. Take a fast 14-inch f/4 Newtonian, for example. It will have a focal length of around 1400 mm. With an astro-video camera having a 1/3 of an inch sensor typical of most entry-level cameras, the resulting image will have a field of view of approximately 8 × 11 arc-minutes. Compare this to the 34 × 51 arc-minute field of view that the same telescope will have with a typical DSLR camera having an APS-C sensor. The difference? With the video camera, an image of the Moon will need a 4×3 mosaic, while a DSLR on the same telescope can easily capture the entire Moon in one frame.

Now take a look at an 11-inch f/10 SCT. It has a focal length of around 2800 mm and a field of view with a 1/3 of an inch video camera of around 4×6 arc-minutes. A 8×5 mosaic is needed to make an image of the Moon. A 0.63x focal reducer gives us a reasonably fast focal ratio of f/6.3. However the focal length is approximately 1800 mm, producing a field of view of 7×9 arc-minutes that requires a 5×4 mosaic to image the Moon.

As you can tell from the example above, the resultant field of view even when using a 0.63x focal reducer remains small. If you use a 0.3x focal reducer, the focal length is reduced to 840 mm, which produces a field of view of approximately 15×22 arc-minutes—still not sufficient to capture the Moon in one frame but more than sufficient for most deep space objects. Even a 150-mm f/10 SCT will have a focal length of around 950 mm and a field of view of 13×17 arc-minutes with a 0.63x focal reducer or a focal length of 750 mm with a field of view 20×26 arc-minutes with a 0.5x focal reducer.

The manufacturer of your camera is a good source to obtain a focal reducer for your telescope that will perform well with your camera. Several companies offer focal reducers ranging between 0.18x and 0.8x. People with astrophotography backgrounds, when they first learn that a 0.18x focal reducer is sometimes used for video astronomy, often think that vignetting will be severe. This is a consideration for traditional astrophotography, but it is not as big a factor for video astronomy. The reason is that the very small sensor used by astro-video cameras typically remain fully illuminated with signal even when extreme focal length reduction is used. The 0.18x focal reducer is camera specific and for Schmidt Cassegrain telescopes having an aperture of 10 inches or more. A variable reducer between 0.29x and 0.64x in six steps is also available.

What is the optimal focal length for video astronomy? There really is none, but 800–1000 mm is often mentioned on Internet forums as the upper limit for 1/3-inch and ½-inch cameras, respectively. From Table 2.4 you can see that the field of view with a 1/3-inch camera and a telescope having an 800-mm focal length is 15×21 arc-minutes, and the field of view with a ½-inch camera with a 1000 mm focal length telescope is 17×22 arc-minutes; both are sufficient to view a large number of deep space objects.

Refractors and Schmidt Cassegrain telescopes are very popular for video astronomy. These telescopes use a star diagonal. One aspect you need to resolve is whether you are going to use a star diagonal with your video camera or not. Most people with experience in video astronomy do not use diagonals with their astro-video cameras. If you do use a diagonal, make sure that the one you are using is collimated. A significant percentage of star diagonals aren't. One way to tell if the star diagonal is collimated is by rotating the diagonal while viewing a star or object with an eyepiece. The star or object should remain stationary as the diagonal rotates. If the diagonal is not collimated and you still want to use it, be sure that you fix the orientation of the star diagonal to the telescope before starting the alignment process, and do not change the diagonal's orientation during the entire session for the night. If you can afford it, a quality star diagonal such as the 1.25-inch Williams dielectric is well worth its price, not only for the optics, which are excellent, but for its mechanical features (collimation, beveled nosepiece, compression eyepiece retention, etc.).

The first steps for starting an observation session with an astro-video camera are the same as for viewing deep space with an eyepiece or photographing it with an astronomical CCD camera or DSLR. First the mount is aligned with the stars, and then, for equatorial mounts, a precise polar alignment is done to align the mount to the celestial pole. Today's equatorial mounts typically have software routines to assist with the precise polar alignment. Although not as accurate as manual polar alignment processes, they are generally more than good enough and save a lot of time.

Regardless of the type of mount used, one aspect of the setup that is very important is an accurate alignment. Use at least a 12.5-mm or shorter focal length reticle eyepiece and precisely center the alignment stars. Unless you have an unusually talented eye, doing it unaided often does not produce the GOTO accuracy desired with a video telescope.

If you want to use your astro-video camera without using a star diagonal you have two alternatives to align the mount to the night sky. If you have a good diagonal that is collimated you can use it to do your alignments and then remove it for your video session. The second alternative is to align the mount without using a diagonal. This means you may need to get down on hands and knees to see through the eyepiece. Here a ground mat such as an exercise mat is very handy to have as well as a very short-legged stool. If you use a poorly collimated diagonal for the mount alignment and then replace it with a camera you will most likely end up spending a significant amount of your observing session acquiring objects in your camera's field of view rather than enjoying deep space in color. *Note:* Some experienced astronomers are capable of doing the alignments using their video cameras. This eliminates most issues caused by either a diagonal or telescope eyepiece.

If you take your time and accurately center your alignment stars and level your mount, objects should appear within the center third of the view screen using a camera having a 1/3 of an inch sensor, an ST80 telescope with a 0.5x focal reducer, and a SLT type mount and tripod. In any case, after completing your mount alignment(s) check the accuracy of your GOTOs. This is simple enough to do and can save you considerable time in the course of the night's viewing session.

Do a GOTO to several objects located in areas of the night sky that contain the objects on your viewing plan for the night. The purpose is to verify the accuracy of your mount's GOTOs. How accurate should the GOTOs be? If the objects are seeable in a 10- to 12-mm eyepiece, then your alignment is excellent. If they are not seeable but close, note the offset and magnitude of the miss. If it is consistent, this will help you locate objects later with the camera installed. Keep in mind that a camera with a 1/3 of an inch sensor produces an image similar to that produced by a 6-mm eyepiece. Once you are satisfied with the mount's alignment, finish by doing a GOTO to the first object on your viewing plan, sync your mount to that object, and then go to a bright star nearby. Center the star in your eyepiece. This star will be used to focus the camera.

If you are using a focal reducer or filter, install it on the 1.25-inch nosepiece of the camera's C mount. Now you are ready for the next step, inserting and focusing the astro-video camera. Inserting the camera is straightforward; just stick it in the drawtube or the diagonal if you are using one. Be careful not to move the telescope and lose the alignment on the bright star. After inserting the camera, be sure to secure the camera so that it does not fall from the telescope.

If you are using a Celestron mount it may have a command called "Calibrate GOTO." If so, select and execute this command; if not, skip the rest of this paragraph. The Calibrate GOTO command will allow the mount to measure and adjust for the impact on the mount's performance caused by replacing the eyepiece with the camera. After the mount is finished with the Calibrate GOTO function, replace the camera with an eyepiece. The bright star most likely will no longer be centered in the eyepiece or may not even be seen in it. Find and center the star. Remove the eyepiece and install the camera again. Be careful not to move the telescope as you remove and replace the camera and eyepiece.

Now that the camera is installed in the telescope and we have a bright star in the center of the field of view, the next step is to focus the camera. This can be a very trying exercise the first time you do it. Set the shutter speed to 1/60 of a second with the automatic gain set to off. At this point, if the camera is significantly out of focus, the image size of the bright star may be so big that the star completely fills the field of view and is not visible on your TV screen. Now slowly turn the telescope's focuser knob until the star comes into view. This will generally start as a very faint, big round dot that slowly focuses to a point of light. Focus on the star, getting the smallest and brightest image possible.

Next, repeat the above step except with the shutter speed set to 1/120 of a second. If the image at 1/60 of a second had several stars in it, you may notice that fewer stars are seeable at 1/120 of a second. Turn the focuser slightly until you maximize the brightness of the stars in the image. Using one of the dimmer stars provides better results than a bright star. Repeat the above using a faster shutter speed until only the bright star you selected is visible and only when focused.

Next go to the camera's "EFFECT" menu and select digital zoom. Enlarge the image to 2x or larger, with larger being better. Now adjust the focus to obtain as small an image as possible on the monitor. The camera is now focused.

The first few times you focus an astro-video camera can be difficult, especially for newcomers to astronomy. But after you have done it a few times, you will know what to expect, and the task becomes far easier to do.

If you have a parafocal ring and a spare 9- or 10-mm eyepiece, the first time you focus your camera you may want to take the opportunity to adjust the parafocal ring on the eyepiece. Once you do this, you can focus using the parafocal eyepiece and then replace the eyepiece with a camera, and the camera will be near focus. All that is required is doing the final adjustment with digital zoom. Many people find this is the easiest method to acquire and compose an image as well as to focus the camera.

If you do not have a parafocal ring and eyepiece, note the position of your focuser when the camera is in focus. With a Newtonian or refractor, you can measure the length of the exposed draw tube. This will give you a place to start the next time you take your video telescope out under the stars. For Cassegrain telescopes, you can count the number of complete turns of the focus knob that is required to either move the mirror all the way forward or backward from focus. Keep in mind, with a Cassegrain telescope moving through the focus area without realizing it is very easy to do. Many find an eyepiece with a parafocal ring makes life much simpler. If you do not have this, you can use electrician's tape to mark the spot on the eyepiece barrel where focus is reached.

A simpler way to focus is to use a Bahtinov mask. If you are handy with working with your hands, you can easily make one. An Internet search using the keywords (Bahtinov mask generator) will turn up many websites that have pattern generators for different-size telescopes that are free for the using. The other alternative is to purchase one. They are inexpensive, around $20–30.

Typically at this point, the first few times you use your astro-video camera you will have wires seemingly everywhere to trip over or get tangled in. The point here is that cable management is very important. Keep wires as short as possible and secured so they do not flop around in the wind, snag on mount protrusions, or, the worst case, are the cause of a telescope tip-over.

The camera will have a cable for power, a shielded coaxial cable to output the video signal, and a cable for the camera's hand controller. The power cable and shielded coaxial video cable can be separate or a twin cable (both cables in one). If you are using a hand controller for your camera, it, too, will have a cable. Find a convenient place on your mount or tripod to store the camera's hand controller when you are not using it. Velcro is an effective solution.

The cables as provided by the camera manufacturers are typically about 25 feet long (7.5 m). This allows for remote viewing but also creates a cable management issue if you are viewing at or very near your telescope. Be creative. Get rid of dangling cables and shorten them where you can, so you don't have wires all over the place. One way to shorten a 25-foot-long cable is to simply fold it until you have the length you want and use a rubber band to secure the folded cable.

A convenient spot for your video display is also needed. The video display should be easily seen from your comfortable viewing chair. Many people prefer a small 7- to 9-inch TV monitor that can be held in your hand and requires little electrical power. Don't forget the GOTO mount's hand controller. Locating it in a position convenient to your viewing chair makes life easier when you want to go to another object or adjust the composition of the image. Again, Velcro is useful. If the mount's hand controller cable is too short for your liking, think about getting an extension cable. Extension cables 25 feet (~7.5 m) and longer work well for many mounts such as the SkyWatcher SynScan AZ GOTO mount and its counterparts from Celestron and Orion.

Once you get the cable issues resolved, now is time to start viewing deep space in near real-time. Now just what is "near real-time?" Near real-time can mean, essentially, viewing bright objects instantly with no time delay or it can mean needing to wait as long as two or three minutes before an image of a dim object is seen. It's this part, waiting a few minutes for an image to appear, that discourages and frustrates some people.

Note The following OSD menus have the AVS DS0-1 camera's OSD menu headings with the MallinCam Micro EX camera's OSD menu headings enclosed in parenthesis. The Revolution Imager is very similar to the MallinCam.

Once the camera is focused, do a GOTO to the first object on your viewing list. Select an integration of X256 with automatic gain set to off. An image should appear on your TV monitor in approximately 5 seconds. If the image is washed out, shorten the integration time until you are satisfied; if it is too dim increase the integration to X512. Don't overlook adjusting the gamma setting in the PROCAMP (ENHANCE) OSD menu as well as the image brightness in the AVSYSTEMS (EXPOSURE) OSD menu. These settings are often dependent upon the object brightness and seeing conditions as well as personal preferences. They can create some challenges for beginners in video astronomy until it is realized that adjusting camera settings is really simple to do if you have a remote hand controller for the camera. Using the buttons on the rear of the camera is also possible but tedious.

Reducing the integration period to improve an image is a concept that often is not understood, as there is a mistaken idea that it reduces the signal level of an image. In a way this is true to some degree. Current budget-level astro-video cameras cannot make an exposure longer than 1/60 of a second NTSC (1/50 of a second PAL). Deep space objects require exposures longer than 1/60 of a second. To get a longer exposure, we stack (add) the signal of two or more frames up to a maximum of 1024 frames. Each frame has the same signal-to-noise ratio, so the sum of 1024 1/60 of a second frames will be 1024 times higher than one 1/60 second frame. However, if the stack of 1024 frames saturates the pixels, then we have too many photons and must reduce the number of frames stacked until we no longer are saturating pixels.

For example, if, say, 1000 photons per second is the saturation point of a pixel and if the signal from a 1/60 of a second frame is equal to say 13 photons per second, a stack of 128 frames provides a signal of 1664 photons per second, and the pixel is saturated. If we reduce the integration to 64 frames we have a signal of 832 photons per second, which does not saturate the pixel. Lowering the number of frames stacked provides the optimum signal-to-noise ratio for this particular example.

Summing up, recall that the stacking routine of an integrating camera adds the value of pixels; it does not average the values or use some other statistical process. If too many frames are integrated, the sum of the pixels will exceed the number of photons a pixel can handle. The pixel becomes saturated, and the excess photons spill over into the surrounding pixels and are wasted. The resultant image will be bright or washed out, with flaring and bloated stars. Reducing the number of frames integrated to a point where none of the pixels are saturated produces a final image that has the optimal signal and noise levels produced by the camera with the existing ambient conditions and equipment used.

M42, the Orion Nebula, with its bright core is a good object to demonstrate the impact of too long an integration period. Figure 6.3 contains three images of M42 with integrations of X32, X64, and X256, respectively. M42 is a difficult object for cameras, as its wide range of brightness generally produces an image with an overexposed core if details of most of the nebula are desired. However, it is a good

Fig. 6.3 M42 image brightness versus integration times

example of what happens when more photons hit a pixel than the pixel can handle. As you can see from Fig. 6.3, as we reduce the integration from X256 (4.3 seconds) down to X32 (0.5 seconds) the flaring in the central core of the nebula is reduced along with the overall details shown in the image. Keep in mind that these images of M42 are what you see on a TV monitor and were made with an 80-mm f/5 refractor. Most experienced observers will agree that the images greatly exceed the tiny smear of gray seen in the eyepiece of an optical 80-mm f/5 refractor.

Note Few objects have as dynamic a range as M42. Photographers use several techniques to overcome the flaring of the central core but keep details at the same time. Generally they take several images with different exposure times and merge them into one using different techniques that are outside the scope of this book.

If you don't like the composition of the image, use the telescope's hand controller to obtain the desired composition. This task is something that experienced astronomers will find easier to do than beginners to astronomy, especially beginners who are on a parallel path trying to learn astronomy and video astronomy at the same time. Composition adjustments can be made with the camera or the parafocal eyepiece.

Once you are satisfied with the composition and image brightness, go to the AVSYSTEMS (EXPOSURE) OSD menu to set the automatic gain to any setting other than off; 36 dB is a good starting value. This enables the 3D-DNR features of the camera. In the PROCAMP (ENHANCE) OSD menu, always set the INTMUL value to 5 and DPC to off. For an integration of X1024, an image will appear after approximately two minutes once you exit the OSD menu and save your changes. If the image is too bright, lower the number of frames integrated until you obtain an acceptable image. If you are having difficulty obtaining a pleasing image, adjust the automatic gain control. Also take a look at the gamma setting in the PROCAMP (ENHANCE) OSD menu. The control has four settings; 1.0, 0.6, 0.45, and 0.3. A setting of 1.0 produces the darkest image, while 0.3 produces the brightest image.

At first, adjusting the camera's settings is often a trying experience. However, most people only need a few nights to become familiar with these and how they relate to each other.

Remember none of the budget-level cameras included in the discussion in this chapter have internal cooling. Thermal noise can be a problem, especially in the summer with its hot, muggy nights. Selecting the DPC option will help some, but the time required to process an image can be several minutes.

If you are using a refractor and stars, especially bright stars, seem bloated on your video display; you may be having an issue with infrared wavelengths, not overexposures or field rotation. With a refractor, infrared light does not focus like the visible light spectrum does. This tends to produce greatly enlarged stars, especially if the star is bright. An inexpensive ultraviolet/infrared filter will restore nice stars to the image. If you still have star bloat with an UV/IR filter installed, you may want to reduce your integration time or take a look at how well your mount deals with field rotation or tracking.

So far in our discussion, we have not mentioned the impact that the type of mount has upon acquiring an image—in other words, the impact of an equatorial mount versus an alt-azimuth mount. Of the two types of mounts, the equatorial mount can adjust for field rotation and produce the sharpest image. In reality, the difference in a near real-time video image produced by either type of mount is often not that significant. This is because the 3D-DNR and stabilization features in current astro-video cameras can adjust for some target movement.

The brighter stars in images made using an alt-azimuth mount are typically a little larger than images of the same stars made with a German equatorial mount. An alt-azimuth mount is more likely to produce round stars, while equatorial mounts often produce smaller but square stars. However there is a limitation regarding the number of integrations that the 3D-DNR feature of the camera can perform while adjusting for field rotation. The camera keeps integrating new images, adding to the signal strength of the image produced. For a precisely polar aligned equatorial mount, the camera can continue its integration for a substantial time until it can no longer adjust for any field rotation present. How long? This really depends upon the mount and its alignment, but ten to twenty minutes is not unrealistic for a quality German equatorial mount. This is not true for an alt-azimuth mount, as field rotation can become noticeable after two or three minutes dependent upon the location of the object in the night sky, which determines the rate of field rotation.

One prime attribute driving the development of video telescopes is the ability to sit inside a warm house observing on your television set while the telescope and mount are located outside in the cold of winter or heat and insects of summer. The 25-foot (7.5-m) cables that come with the budget-level cameras are rather short and can be extended considerably. How far? Based upon the performance of a CCTV camera, distances between 700 and 800 feet (210–240 m) are possible, depending on the quality of the RG 59 coaxial cable used.

Two types of RG 59 coaxial cables are available. The highest quality has a center wire that is 100% pure copper with a 95% braided copper shield. A cheaper RG 59 coaxial cable typically used for home television installations has a copper-plated steel center wire and a braided shield made of aluminum. Another coaxial cable, RG 6, can be used. It has larger diameter center cable and can support distances up to 1000 feet (300 m).

Keep in mind that the longer the cable, the greater the degradation of the signal. If you are considering distances longer than 100 feet or so (~30 m), visit CCTV installation websites. Several have excellent tutorials regarding cable runs. Do an Internet search for "maximum length of a video cable from a camera to a TV monitor."

Locating a TV monitor 100 feet from a telescope with the camera is relatively easy to do. However, the same is not true for controlling the telescope. Two basic options exist. One option is to make an extension cable for the mount's hand controller. Another option is to control the telescope mount with a computer. This can be done with hard wires or wireless.

Functional extension cables for Celestron hand controllers have reportedly (on Internet forums) been made for up to 100-foot distances (~30-m). You will need to align your telescope and focus your camera outside, but once that is completed you can move inside and view the night sky in comfort. You will also need a cable for the camera's remote control. Once you have the video signal inside your home or viewing shelter, you can connect it to a TV monitor, large screen television, overhead projector, DVR, or even a computer. Once you connect to a computer, you can send your video worldwide via one of several broadcasting sites or use your video telescope for astrophotography.

Using a computer to remotely control your telescope is more complex than having simple extension cables for your camera and telescope's hand controller. However, the computer makes up for this complexity by offering greater capabilities. Perhaps the easiest method to remotely control a telescope is a wireless setup such as SkyFi for IPads/IPhones or SkyBT BlueTooth telescope control for

Skysafari on Android tablets and smart phones. In electronically quiet zones ranges up to 100 m are sometimes possible. If other wireless networks, etc., are present then the range is shorter.

Maksutov Cassegrain Telescopes and Video Astronomy

Small Maksutov Cassegrain telescopes are very popular due to their compact size and images. Unfortunately these telescopes typically have very long focal lengths. The question often arises whether a small Maksutov Cassegrain telescope be used for video astronomy. The answer is yes, but there are some limitations.

Life with a Maksutov Cassegrain telescope and video astronomy is much easier if the rear threads of the Maksutov Cassegrain telescope are converted to the standard 2-inch rear ones used by Schmidt Cassegrain telescopes. All that is needed is an inexpensive adapter to convert the rear threads (see Fig. 6.4) and a 1.25-inch SCT visual back.

Maksutov Cassegrain telescopes have either a male 35-mm or 44.5-mm rear thread. The 35-mm rear thread is used by Meade's ETX series Maksutov Cassegrain telescopes and the Celestron 4SE. The 44.5-mm rear thread is used by Celestron, Orion, and SkyWatcher for their 127-mm Maksutov Cassegrain telescopes, some 90-mm telescopes, as well as for their 150-, 180-, and 190-mm Maksutov Cassegrain telescopes.

Since manufacturers can change their specifications, check the threads on your telescope before ordering an adapter. The 35 mm is slightly larger than 1 3/8 inches; 44.5 mm is slightly larger than 1 3/4 inches. If your Maksutov Cassegrain telescope

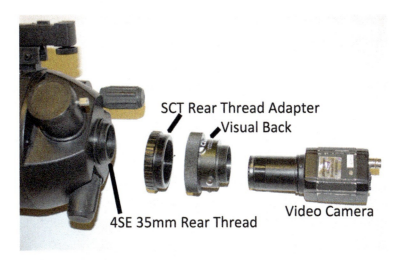

Fig. 6.4 Maksutov to Schmidt Cassegrain adapter

Fig. 6.5 4SE telescope on a wedge with a video camera

has 35-mm rear threads you will need the Baader SCT Adapter part number 2958500A. If you have a telescope with the 44.5-mm rear threads then you will need the ScopStuff SCT adapter part SMSA.

Once you have converted the rear threads to the standard 2-inch Schmidt Cassegrain rear threads, add a 1.25-inch visual back, Celestron part number 93653-A (see Fig. 6.4). Now you can easily use many of the 1.25-inch accessories available for the Schmidt Cassegrain telescopes, including your video camera.

To use a video camera, all that you need do is attach the adapter to the rear of your Maksutov, screw on the visual back, and then insert your camera. See Fig. 6.5 for an example with the Celestron 4SE. This popular telescope comes with a wedge, allowing operation as either an alt-azimuth or as an equatorial telescope.

Two focal reducers are easily used with Maksutov Cassegrain telescopes: the SCT 0.63x focal reducer and a 1.25-inch 0.5x focal reducer. Here is the process for attaching the ST 0.63x focal reducer to the telescope:

1. The SCT rear thread adapter is screwed directly onto the rear threads of the Maksutov Cassegrain telescope.
2. The 0.63x SCT focal reducer is screwed directly onto the SCT rear thread adapter.
3. The 1.25-inch visual back is screwed onto the focal reducer.
4. The 1.25-inch C-mount is screwed into the camera and then inserted into the 1.25-inch visual back.

The process for using the 1.25-inch 0.5x focal reducer is similar:

1. The SCT rear thread adapter is screwed directly onto the rear threads of the Maksutov Cassegrain telescope (Fig. 6.6).

Fig. 6.6 An SCT focal reducer attachment for a Maksutov Cassegrain telescope

Fig. 6.7 A 1.25-inch focal reducer attachment for a Maksutov Cassegrain telescope

2. The 1.25-inch visual back is screwed onto the SCT rear thread adapter.
3. The 1.25-inch 0.5x focal reducer is screwed into the 1.25-inch C mount.
4. The 1.25-inch C-mount is screwed into the camera and then inserted into the 1.25-inch visual back.

As shown by Fig. 6.7, a 1.25 inch 0.5x focal reducer screws directly onto the video camera's 1.25-inch to C-mount nosepiece adapter which is inserted into the 1.25-inch SCT visual back. Further reduction in focal length may be obtainable by using both focal reducers in tandem. How effective the combination of the two focal reducers will be is questionable but certainly worth trying.

Table 6.1 Maksutov Cassegrain telescope focal lengths

Aperture (mm)	Focal length (mm)	Focal length: 0.63x focal reducer (mm)	Focal length: 0.5x focal reducer (mm)
90	1250	790	625
102	1325	835	660
125	1900	1200	950
127	1500	945	750
150	1800	1135	900

The 4SE telescope has a focal length of 1325 mm. This will produce a small field of view and dim image. The 0.63x SCT focal reducer will shorten the focal length to 835 mm, while the 1.25-inch 0.5x focal reducer will shorten it to 665 mm. The 665-mm focal length will work very well for the brighter deep space objects. It will also be more forgiving with vibration and tracking movements. The option to use either an SCT 0.63x focal reducer or a 1.25-inch 0.5x focal reducer is applicable to most small aperture Maksutov Cassegrain telescopes. Table 6.1 shows the focal lengths for some popular Maksutov Cassegrain telescopes without a focal reducer, with a 0.63x focal reducer, and with a 0.5x focal reducer. As shown in the table, even with a focal reducer, the focal lengths of some of the Maksutov Cassegrain telescopes remain fairly long, which will complicate their usage for video astronomy. Only the small 90-mm and 102-mm Maksutov Cassegrain telescopes have focal lengths that are really supportive of video astronomy. This is not to say that you cannot use the larger aperture Maksutovs for video astronomy, but you will have issues with small field of view, dim images, and sensitivity to mount movement and vibration.

Chapter 7

Imaging the Night Sky

The Versatility of Video Imaging

One very logical question many people ask is, "Can I use my astro-video camera to photograph the night sky?" The answer is yes, you can make some nice images, some suitable for framing and hanging on your office or den wall. However, they will not compete with images produced with a DSLR or an astronomical CCD camera.

The imaging process with an astro-video camera is not as straightforward as that associated with a DSLR. Several options are available, from something as simple as doing a screen capture to as complicated as stacking a few thousand images in a process very similar to that used by short exposure astrophotography. Imaging with an astro-video camera is also very dependent upon the computer software used to capture, process, and store the video files in a computer as well as the telescope and mount used.

Even though the photographic processes used with astro-video cameras are not as well defined as traditional astrophotography the basic process flow is straightforward. The analog signal from the video camera is changed into a digital signal that is stored and processed in a computer and then printed on paper or shared digitally with others (see Fig. 7.1).

Another aspect regarding astrophotography with an astro-video camera is the resolution of the images produced. As discussed earlier, astro-video cameras have large pixels and small sensors. Large pixels help provide the sensitivity needed to image faint deep space objects while the small sensor size restricts the number of pixels that can be used. Large pixels in a small sensor produce a camera with low resolution, at least in comparison to cameras typically used for astrophotography.

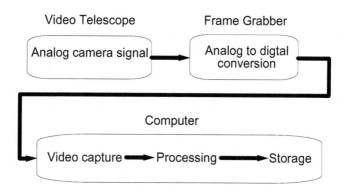

Fig. 7.1 Imaging work flow

Although low resolution is not much of an issue for viewing, it is for producing photographic images rich in details and contrast. Decent images of a wide variety of deep space objects are possible with an astro-video camera, but they generally do not approach the clarity, contrast, and details obtainable from a dedicated CCD astronomical camera, a DSLR camera, or a micro 4/3 mirrorless camera. The relatively low resolution produced is not conducive for enlarging images.

Modern entry-level astro-video cameras have significant in-camera image processing capabilities. Recall that they are making 60 images per second for an NTSC camera (50 images per second PAL). These cameras are integrating cameras and can combine (integrate) an image by linearly stacking up to 1024 of the 1/60 second images. This can produce an image having the same signal-to-noise ratio as one 17-second exposure (20 seconds for PAL). In addition to the integrating capability, modern cameras also have digital noise reduction features known as 3D digital noise reduction in the CCTV industry. This feature further enhances the camera's image by removing motion blur, dark frame subtraction to remove hot pixels, and noise reduction by stacking up to five of the integrated exposures. Depending on the number of images integrated, the final image produced can have the same signal-to-noise ratio of one 85-second exposure NTSC (100 seconds PAL).

These characteristics should make an astro-video camera very suitable for very short exposure astrophotography, which is based upon stacking camera exposures of 30 seconds or less to produce a final image. However, the astro-video camera has a couple of drawbacks related to use as a camera for astrophotography. The first is that the resolution of its sensor is low in comparison to cameras traditionally used for astrophotography. The second is that it has an analog output that is converted into an 8-bit digital signal.

What camera features to use and how to use them depends upon the equipment chosen to image the deep space object as well as what computer programs are used. The variety of computer programs available range from generic video programs for making family movies to programs specialized for astronomy. The way each program operates and how it processes data as well as interacts with other programs varies considerably. The result is that the processes used for imaging with an astro-video camera are numerous and confusing to many. Regardless of the computer program(s) and the equipment used, as shown earlier in Fig. 7.1, imaging with an astro-video camera is straightforward.

Analog-to-Digital Conversion

Recall that the output from an astro-video camera is an interlaced analog television signal with a frame refresh rate of 30 (NTSC) or 25 (PAL) frames per second, complying with either the NTSC or PAL standard. Since it is not a digital signal, we cannot feed it directly into a computer for processing. First we must convert the analog signal into a digital signal. Once the analog television signal is digitized, the computing power of the computer is available to process, enhance, and store the images captured by the astro-video camera. These images may be in the form of a video or a photograph of an object. In any case, the analog to digital conversion used to convert analog video signal into a digital signal is an 8-bit process.

Converting an analog television signal into a digital video signal is not difficult. This task is done with the video capture device called a frame grabber. For the very popular laptop computer an external USB device is used (see Fig 7.2). Desktop computers use either an external USB device or a PCI card installed in the computer. Regardless of whether an external USB device or a PCI card is used the process is the same.

One of the major reasons astro-video astronomy is popular is the ability to have the telescope outside while viewing inside with all the comforts of home. The analog television signal from an astro-video camera can easily be transmitted over 100 feet (30 m), while a USB digital signal is limited to a very short distance.

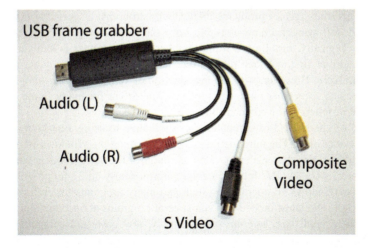

Fig. 7.2 USB frame grabber

Fig. 7.3 Frame grabber connection to a laptop computer

For this reason the frame grabber is located at or near the computer. Also the analog signal can be directly connected to a large-screen television for visually viewing an object. The typical hookup will be the astro-video camera, television cable, frame grabber, and the computer. So we have the composite analog signal from the astro-video camera connected to the frame grabber by a long video cable and the frame grabber plugged directly into the computer, as shown in Fig. 7.3.

The computer industry, in the opinion of many, thrives upon planned obsolescence. This often creates problems in amateur astronomy, as perfectly good hardware and soft ware become unusable with the latest computers and operating systems. Typically, installing a frame grabber on a computer is straightforward. Load the driver first and then plug in the frame grabber. The computer will take over from there. However, with the introduction of Windows 8, this simple installation became somewhat more complicated. The driver for many frame grabbers had to be installed in the compatibility mode.

While on the subject of computers, an alternative to using a desktop or laptop computer does exist. The introduction of tablets running the Windows 10 operating system adds a new dimension to video astronomy. The latest models with 64-bit Quad Core 1.44/1.84 GHz Intel processor can run nearly all 32-bit programs written for the Windows 10 operating system and many programs for earlier versions of Windows. The latest models can run several programs at once without bogging down. Their small size, internal power supply, low cost, and computing power make them ideal in situations where portability is desired or needed. Chapter 9 has additional information Windows 10 tablets and video telescopes.

Digital Processing Overview

Now that the video signal is inside the computer it can be processed and stored. Perhaps the simplest processing and storage is live view. This essentially uses the computer as a television monitor to visually view the night sky real-time—near real-time for the purist, as a small time lag due to camera processing does exist. For this method, the video capture program being used needs to produce a live view display. None of the signal is saved. The camera's settings, such as the number of images integrated, exposure times, brightness, 3D DNR, automatic gain control, etc., are adjusted to produce the image desired.

Dependent upon the computer program being used, a simple command ("snapshot," "single image," "single frame," etc.) will have the computer capture and store the image displayed on the computer screen. If the computer program does not have the ability to capture a single image, then the computer's "Screen Capture" command will do the job. Unfortunately the Screen Capture command generally contains parts of the screen not related to the image. If this is the case, the non-image parts of the screen save are removed using a photo processing program. A photo processing program can also enhance the captured image somewhat, but the 8-bit nature of the digital signal limits the scope of the enhancements that are doable. Again, though, this process, live view with a screen capture, is perhaps the simplest way to capture and save an image of an object in the night sky using an astro-video camera. It also places very few demands upon computing power and data storage.

Video editing programs can store the live view digital signal from the camera as a video file for future viewing or processing. First, adjust the camera's parameters until the displayed live view image is satisfactory. The video capture program will have a command such as "Start Capture, Capture Video, etc." that will start the capture process. The digital signal from the frame grabber is saved in one of the popular video formats. This process can be very intensive for storage of media, as video files often take substantial amounts of memory.

There are several methods to produce an image from a stored video file. One technically simple but labor intensive approach is to convert each frame of the video file into a photo format such as JPG or BMP, visually look at each frame, and then select the best frame as the photographic image of the object. Deciding which image is the best is difficult, as often most of them will look pretty much the same. Once the final image is produced and saved, the large video file can be discarded to save memory space.

Up to this point in our discussion, only techniques doable using generic video and photo processing programs have been discussed. However, many programs exist that are specifically designed for astronomy and astro-video cameras. These programs can improve the photographic process and the results obtained.

Several stacking programs exist for astronomy. Once a video file is stored and converted into a photographic file, a stacking program can stack each frame of the video file. The stacked image will have a better signal-to-noise ratio than any one of the frames that were stacked. Dependent upon the number of frames stacked and the noise in each frame, the final stacked image may even have a bit of depth greater than 8 bits. The stacked image is enhanced using a photo processing program. However, the magnitude of the enhancement is limited due to the 8-bit nature of the converted analog signal. After the final stacked image is produced, erasing the video files frees up the computer memory for another day.

SharpCap is an interesting and currently very popular freeware program designed specifically for video astronomy. It can operate as a traditional screen capture program. However, the feature that sets it apart is its ability to stack real-time each frame of a live video feed and also adjust for field rotation at the same time. This allows sharp images from video telescopes using alt-azimuth mounts. By stacking each frame as it comes into the computer, storage memory is not impacted, noise is greatly reduced, and color depth increases dependent upon the number and noise characteristics of the frames being stacked. Changes in the image (integration time, brightness, etc.) at least for entry-level astro-video cameras, are made via the camera's OSD adjustments or by adjusting the image's histogram. One option allows saving only the stacked image or images, which greatly reduces demands upon storage media. The final product can also be enhanced in photo processing programs such as *Photoshop®*, producing images of the brighter deep space objects that can compete in some aspects with traditional astrophotographs of the same objects.

Setting Up for Astrophotography with an Astro-Video Camera

As with imaging using an astronomical CCD camera or a DSLR camera, the photographic session for the night should be planned ahead of time. The combination of telescopes having relatively long focal lengths with video cameras and their small sensors produces images having a small field of view (large image size). This is a major characteristic to consider while selecting which deep space object will be the target of an imaging session.

A good telescope for a beginner to astronomy, including video astronomy, is an 80-mm f/5 refractor on an entry-level GOTO telescope mount with a camera having a 1/3-inch sensor. Just about any GOTO alt-azimuth mount will do. iOptron, Orion, and SkyWatcher sell GOTO alt-azimuth mounts without telescopes, while Meade and Celestron bundle their GOTO alt-azimuth mounts with their telescopes. The combination of an 80-mm f/5 refractor with a camera having a 1/3-inch sensor has a field of view of 41×31 arc-minutes. A 0.5X focal reducer is inexpensive and is easy to install. It will increase the field of view to 83×62 arc-minutes. This relatively large field of view makes both finding and then tracking objects much easier than doing the same task using a telescope having a longer focal length, especially for telescope focal lengths greater than 800–1000 mm.

The setup for astrophotography with an astro-video camera is much the same as for a visual session except a frame grabber and a computer are needed.

However, planning a photographic session is much more detailed and important than planning for a visual session. Consideration must be given to:

- the object(s) you plan to image
- the characteristics of the object you want to image
- the location of objects in the sky during the time you plan to image
- the time of night that you will be imaging the objects.

Obviously, you will need to decide which objects you want to image. Once you have decided upon your candidate objects, you will need to determine if your camera and telescope are up to the task to image them. Will the object be too big or too small for your astro-video telescope? Will it be bright enough or too dim to image?

For example, let's assume that we have an 80-mm, f/5, short tube refractor telescope and a typical entry-level astro-video camera having a 1/3rd-inch sensor. Assume that the image we want to photograph is M45, the Pleiades. An 80-mm, f/5, short tube refractor will have a focal length of 400 mm. Recall our discussions about field of view in Chap. 2. From Table 2.3 we can see that a camera with a 1/3-inch sensor connected to a telescope having a focal length of 400 mm will have a field of view of 41×31 arc-minutes. Next we look at the apparent diameter of M45 and find that it is 120 arc-minutes. This is about three times larger than the field of view produced by an 80-mm, f/5, telescope with a 1/3-inch camera. If we use a 0.5x focal reducer with the telescope, we will shorten the focal length to 200 mm and increase the camera's field of view to 83×62 arc-minutes. Although this is not sufficient to capture M45 in its entirety, it is sufficient to capture most of the object and may produce a final image that is pleasing to the eye.

Now consider the same situation with the ever-popular 8-inch SCT. This telescope will have a focal length of 2000 mm and a focal ratio of f/10. From Table 2.3 we can see that the field of view of an 8-inch SCT with a camera having a 1/3-inch sensor will be 8×6 arc minutes. Recall that the apparent diameter of M45 is 120 arc-minutes — about 12 times larger than the field of view produced by the 8-inch SCT with a 1/3-inch video camera. Using the standard 0.63x SCT focal reducer will increase the field of view in the video camera to around 13×10 arc-minutes. This is about 8 times smaller than the apparent diameter of M45 and is not likely to produce a pleasing image. Adding a 0.5x focal reducer in tandem with the 0.63x focal reducer will shorten the telescope's focal length to 630 mm and result in a field of view around 28×21 arc-minutes, which is about 1/3 the apparent diameter of M45. Again, this is not likely to produce a pleasing image. As discussed, even the field of view of an 80-mm refractor with its focal length shortened to 200 mm using a 1/3-inch sensor was not large enough to capture all of M45. The conclusion is that large objects such as M45 make difficult targets for video telescopes.

Lest we become too pessimistic, most deep space objects are far from Earth and have small apparent sizes. Large objects such as the Pleiades, M45, and M31 are the exception. Nearly all of the 110 objects on the Messier list have a small apparent size. The fields of views of either an 80-mm f/5 refractor with a 0.5x focal reducer or a 2000-mm f/10 SCT with a 0.5x and a 0.63x focal reducer in tandem are adequate for imaging all but a small handful of the Messier objects.

After determining that our telescope and camera can in fact capture the objects we wish to image in their field of view, the next step in our session planning is to see where the objects are in the night sky. A computer planetarium program such as *Stellarium* is ideal for this task. The objectives of this part of planning the imaging session are to determine:

- if the object is visible in the night sky
- that no houses, trees, etc., block the view of the object
- if and when the object crosses the meridian
- if the object tracks near the zenith.

It is obvious that the object must be in the night sky to photograph it. Here, however, the interest is where in the sky the object is located at the start of the imaging session and where it is located at the end of the session. Also, will any objects such as tall buildings, trees, etc., obstruct the object as it tracks through the night sky? If the object is in the western sky, is it high enough to allow sufficient time for the imaging session before it gets too low on the horizon? If it is in the eastern sky and you are using a German equatorial mount, will the object cross the meridian during the imaging session? If you are using an alt-azimuth mount, keep in mind the closer the object gets to the meridian, as well as to the zenith, the greater the impact of field rotation becomes.

Next are the logistics-related items. Make sure you have:

- fully charged batteries
- clean lens and mirror surfaces
- all components present and in working order
- all cables present.

Sometimes it's easy to overlook topping off the charge of your battery pack(s). A poorly charged battery can run out of juice in the middle of an imaging session, leaving you out of business for the night. Dirty optical surfaces can certainly ruin a photographic session. However, don't overdo cleaning optical surfaces; take the extra time to keep them clean in the first place. Most valuable is a checklist that has all of your astronomy kit listed. Use it to be sure you haven't forgotten an item. This is especially important if you are traveling to a viewing site away from your home. Don't forget the cables. Video astronomy has a lot of cables, as do auxiliary devices such as computers and tablets with their OTG cables, USB cables, etc. Having a cable list separate from the equipment list can help insure that an important cable does not get left behind.

Resolving these issues before you go out for the night can be the difference between a fun night shooting the stars and a frustrating night where nothing seems to go right. Planning will help produce satisfactory images, especially with an entry-level video astronomy kit. Although the images obtained may not be images that will win a photo of the year award in some astronomy magazine, they will be good enough to proudly show family and friends.

Another issue related to planning remains to consider. Recall the very small field of view associated with telescopes having focal lengths that are greater than around

500 mm (with or without a focal reducer). This small field of view often makes finding objects very difficult for many people, especially people who are not very knowledgeable of star patterns in the night sky. This difficulty of finding objects is increased for people using inexpensive alt-azimuth GOTO mounts that are not as precise as more expensive alt-azimuth or German equatorial mounts. Spending what seems like hours trying to get an object in the field of view of a video camera is a very frustrating experience. This is especially true for a very dim object that requires medium to long integrations to be visible on a TV or computer monitor.

Many people experience difficulty finding objects with an astro-video telescope, whether it is for photography or near real-time viewing. A workaround is a parafocal ring and an eyepiece in the 10-mm range that is dedicated to the task of finding objects in the night sky (see Fig. 7.4). Since the eyepiece is used only for finding objects, image quality is not important, so an inexpensive eyepiece will do. Adjust the parafocal ring so when you put the eyepiece in your telescope, it will be at the video camera's focus. Do your GOTOs using the parafocal eyepiece. Once you have the object centered in the eyepiece, remove the eyepiece and insert the astro-video camera.

The trick here is getting a good focus with the video camera. Using the eyepiece you normally use for GOTOs, locate a bright star near the first object on your list for the night using the GOTO. Once there, center the star in the eyepiece. Next insert the video camera and focus it on the star. A Bahtinov mask greatly facilitates focusing. Once you have the video camera focused, remove it and insert the parafocal eyepiece. Never adjust the telescope's focuser when the parafocal eyepiece is being used, as the focus is set for the video camera. Adjust the parafocal ring until the star is focused in the eyepiece. During the course of a viewing or imaging session resist the urge to tweak the focus of the parafocal eyepiece. All you need the parafocal eyepiece to do is find and center objects in the telescope's field of view. It is not for viewing.

Fig. 7.4 Parafocal ring on an eyepiece

Computer Programs Used in Astro-Video Photography

Probably the simplest way to make an image of a deep space object is to connect an astro-video camera to a computer for viewing. The camera's analog video output is digitized with a frame grabber and fed into a computer, where a video editing or capture program displays the image on the computer screen in what is called "live view." This setup is essentially the same as the near real-time view we saw using the astro-video camera and telescope for viewing the night sky with a TV or LCD monitor in Chap. 6. The only difference is that now we are digitizing the camera's output and are using a computer screen for viewing.

Just about any video editing or capture program will work whether it is a basic program for home videos or specialized for astrophotography. However, not all have a live-view capability. Many frame grabbers are bundled with *Ulead Visual Studio* or *AM Cap*. These programs are for home videos but are simple to use, freeware, and have a live-view capability. Many astronomy programs are also capable of a live-view display. One excellent astronomy program, *Gstar4*, has a window in the screen for the live-view image. The astronomy program *Sharp Cap* also produces a full screen display.

The live-view function greatly enhances the ability to adjust the camera's settings to obtain the best image from the camera. Keep in mind that the camera operates internally with a 16-bit environment, while the frame grabber produces 8-bit images. For this reason it is best to do as much image enhancement with the camera as possible rather than doing it later with a photo editing program. Adjustments for the camera are the same as were used for near real-time visual viewing. Some good starting points for camera settings are:

- AGC = 36 dB
- INTMUL = 5
- STABALIZER = On. The zoom function is disabled when the stabilizer function is selected.
- SHARPNESS = 20–25
- GAMMA = 1. If more details are desired, change to 0.3–0.45.
- INTEGRATION: Adjust until the image is only slightly overexposed. X256 is a good starting point for most deep space objects.
- BRIGHTNESS = 30–99. Gradually reduce from 99 to a value that produces a satisfactory image. Do not use brightness under 30; instead shorten the integration by one step.

The 3D DNR feature can be disabled by setting INTMUL = 0, as it is not needed to adjust the camera. An INTMUL setting of 5 will essentially cut noise in half but will increase the camera's processing time by a factor of 5. When you finish adjusting the camera set INTMUL back to 5.

Now that you have the best image you can acquire on your computer screen, use your computer's "Alt-PrintScreen" command and do a screen capture. This will capture the image on the computer monitor and store it in your computer. The resolution of the image will equal that of the computer monitor. Later you can use a photo processing program to crop out any areas of the image showing the computer screen's display. You can also enhance the image to some extent, but in general what you see on the screen is what you get. This process is fast and uses relatively little computer memory. Keep in mind that not all screen capture programs are equal. Features such as automatic file naming, saving, and numbering captures in a selected folder simplify managing your data and images. *Corel CAPTURE* is an example of a good screen capture program.

A variation of this method is to use your computer program to capture and store one frame of the video output from the camera. Since only one frame is captured, little demand is placed upon the computer's memory. Again, a photo processing program is then used to enhance the image, but since it is an 8-bit image, little room is available for much improvement. Most, not all, computer programs used for video capture will allow the capture of one frame that is stored as a photo. This method is simple and can produce excellent results. It is the preferred imaging method for many video astronomers. A popular variation is to capture several frames over a period of time and then use the best frame for the final image.

The image here of an open star cluster (M36) (see Fig. 7.5) was made using the program *Visual Lead* to capture one frame. The camera settings were:

- Integration: X16
- AGC: 24
- 3D-DNR: INTMUL = 5
- Telescope: f/5 ST80A
- Mount: 4SE Equatorial Mode.

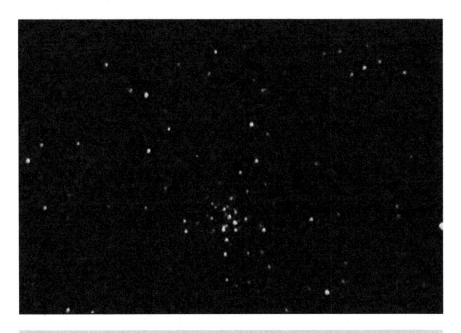

Fig. 7.5 M36

Fig. 7.6 M17

M36 is located in the Constellation Auriga. It has a magnitude of 6.0 and an apparent diameter of 12 minutes. The cluster is 4300 light years from Earth.

The image of the Swan or Omega Nebula (M17) (see Fig. 7.6) is another example of one made by the program *Visual Lead* to capture a single frame. The camera settings were:

- Integration: X1024
- AGC: 0
- 3D-DNR: off
- Telescope: f/5 ST80A
- Mount: 4SE Equatorial Mode.

M17 is located in the constellation Sagittarius. It has a surface brightness of 21 mag per arc-sec^2 and an apparent diameter of 40×30 minutes. The cluster is 5900 light years from Earth.

Another method used to produce images from an astro-video camera is very similar to that used for traditional astrophotography. The photographer takes many images of the same object and then stacks them to produce a final image. This process can produce a higher quality image than the screen capture process discussed in the previous section.

For the traditional stacking process used to produce an image of an object, a video is made of the object over a period of time. The video is stored, and then the best quality images are selected for stacking. A traditional stacking program such as *Deep Sky Stacker* or *Registax* is used to stack the images. The final step is using a photo processing program to enhance the image.

A major issue associated with stacking images to produce a final image is the large amounts of computer memory required for storing video and photographic files and the limitations associated with enhancing 8-bit images. Also, some stacking programs cannot stack AVI files. So each frame of the AVI file must be converted into a bitmap file.

Also keep in mind that the frame rate of a video camera and the frame capture time of a video capture program are not the same thing. Recall from Chap. 2 that the video camera will refresh the image it produces at either 25 or 30 frames per second depending on whether it is a PAL or an NSTC standard camera. Most images of deep space objects need integrations around X128 or greater. An integration of X128 takes 2.13 seconds without 3D-DNR selected. During this time a NTSC standard camera will refresh its image 64 times. The camera will send 64 identical images to the television monitor while the camera is integrating 128 frames. In other words, for every unique image produced by the camera from an X128 integration, 64 identical images are sent from the camera to the computer.

This creates an issue related to stacking images. Again, recall from Chap. 2 that stacking the same (identical) image essentially does nothing to improve the signal-to-noise ratio of an image. If we want to improve this ratio, say, by a factor of ten, 100 unique images must be stacked. In the example above for an X128 integration, 213 seconds are needed to produce images made from 100 X128 integrations. During this time the camera will have sent 6390 images to the computer, of which only 100 are unique. The stacking program will ignore the identical images, but they will take up both processing time as well as computer memory. In this example, the impact of the camera's 3D-DNR feature was not included. If they are included, then the times and number of images involved increase by a factor of five.

Three options are available to work around this issue. One option is simply to ignore it and co-exist with the memory needs and time required processing the large number of images involved. The second is to manually grab an image during each integration period, something that is tedious to do. The third is to use a computer program to select an image produced by each integration. Obviously, the third option produces the best results.

The freeware capture program *GStar4Capture* has an adjustable capture rate. It can be set to capture one frame during an integration. The program, while excellent, is getting a bit dated and can only accommodate integrations without 3D-DNR that do not exceed X256. Another feature of the program that is exceedingly useful is its ability to change frames from an AVI video file into one of several popular bitmap image formats. Since the AVI frames are in sequential order, after conversion to a bitmap file, the images can be reviewed and one image from each integration period selected for stacking (with or without 3D-DNR). Once the images needed for stacking are selected they are imported into a stacking program and stacked. Which stacking program is the question.

Astronomy Video Capture and Stacking Programs

Several computer programs are available for capturing the signal from an astro-video camera and then saving it for processing later. Some even include the ability to stack and process images. Which one is best? That is difficult to say. Computer programs tend to be personal. What one person thinks is the best computer program ever made another thinks is trash. Also the needs and equipment used for video astronomy varies considerably among amateur astronomers. Here are some of the more popular computer programs used by amateurs in video astronomy community:

- *SharpCap*: http://www.sharpcap.co.uk/
- *Deep Sky Stacker:* http://deepskystacker.free.fr/english/index.html
- *RegiStax 6:* http://www.astronomie.be/registax/
- *AutoStakkert!2:* http://www.autostakkert.com/wp/
- *Astro Live:* http://astroprecisioninstruments.com/
- *Gstar4Capture:* http://www.astroshop.com/

SharpCap is a very popular astronomy video capture and stacking program that, thanks to its developer, is freeware. This program, specifically designed for video astronomy, has several capture options:

- *Snapshot* is just what it implies. It makes a PNG or FITS image of the image displayed on the screen.
- *Quick capture* makes an AVI video containing a predefined number or frames or for a predefined length of time.
- *Capture* is a manually controlled version of the quick capture feature.
- *Live stack* continuously stacks each image as it is fed to the computer and saves only the stacked image. The feature has the capability to adjust for field rotation as well as for telescope movements. It requires an excellent focus to detect the stars required to adjust for field rotation and telescope movements.

SharpCapture has several other features that are very useful, such as adjustments to the stacked image's histogram and focusing support (Bahtinov mask, full width at half maximum (FWHM) measurements of stars in the image). Further information and downloads are at: http://www.sharpcap.co.uk/sharpcap.

DeepSkyStacker is one of the more popular stacking programs for amateur astrophotographers. Thanks to the generosity of its developer it is freeware and available in many languages. The program is designed specifically for stacking images of deep space objects and comets. However several video astronomy-related issues exist with *DeepSkyStacker*. These are:

- It has no video capture capability, so a separate video capture program must be used to capture and store video files in a computer.
- It cannot process video files. Each frame of a video file must be converted into a TIFF, FITS, JPEG, BMP, PNG, or a DSLR RAW file.

- The program requires at least eight stars common to all frames that it stacks. Due to the small field of view associated with astro video cameras, finding eight common stars is sometimes difficult.
- In spite of the program's vast array of advanced capabilities and features, which are far too numerous to discuss here, it is easy to learn and to use. The Patrick Moore Practical Astronomy book *Astrophotography on the Go* has an excellent tutorial for the program. Further information and downloads are available at: http://deepskystacker.free.fr/english/index.html.

RegiStax 6 is a mature stacking program containing sophisticated image-enhancement features. It is designed specifically for stacking images from AVI files and is the program of choice by many video astronomers. *RegiStax* is particularly suited for Solar System images. Thanks to the generosity of its developer, it is offered as freeware. For additional information: http://www.astronomie.be/registax/.

AutoStakkert!2 is a stacking program for video files. It has the ability to adjust for field rotation, which makes it very handy for people using alt-azimuth mounts. Like *RegiStax*, it has some sophisticated image enhancement features. Many video astronomers like to stack with AutoStakkert, then use *RegiStax* to enhance the image. It, too, thanks to the generosity of its developer, is freeware. Additional information about the program and its capabilities is located at: http://www.autostakkert.com/wp/.

Astrolive by Astro Precision Instruments LLC is a program covering both video capture and processing. At this time it costs $99 and is available for a free 30-day trial. Among its long list of features are the ability to control GOTO telescope mounts, framing and focusing, live view, live stacking, field de-rotation, histogram adjustment, high definition video capture, and camera control, just to name a few. The program has an intuitive interface and a relatively easy learning curve. For detailed information and downloads see: http://astroprecisioninstruments.com/.

Gstar4Capture is an astronomical video capture program that is freely available thanks to its developers. It has video capture rates for integrations between X2 and X256, making it very useful as a companion to stacking programs such as *DeepSkyStacker*. *Gstar4Capture* has a long list of features that make it a very powerful and useful video capture program. A partial list of these features includes translation of video files into bitmaps, night view mode, on-screen zoom, occultation time stamping, automated script and manual capture modes, guiding reticles, time lapse, and image enhancements. Since the AVI frames are in sequential order, after conversion to a bitmap file the images can be reviewed and one image from each integration period selected for stacking (with or without 3D-DNR). As the authors say, it is a complete video capture tool for astronomy. For additional information and download see: http://www.astroshop.com/.

Astrophotographers will notice that several aspects of stacking are not discussed, such as dark, flat and offset/bias frames. This was done to keep the topic of imaging with an astro-video camera a basic as possible in order not to overwhelm a beginner to both astronomy and imaging.

Fig. 7.7 Messier object 3

Dark frames are really not needed, as the camera has the ability to do dark frame subtraction as part of its 3D-DNR feature. Due to the rather small field of view associated with astro-video cameras, corrections for vignetting are typically not required, as the sensors generally remain fully painted even when strong focal reducers are use. Dust motes, on the other hand, are often a problem. Dark, flat, and offset/bias frames can be used if desired. The processes for acquiring and using them are very similar to that used for astronomical CCD cameras and DSLR cameras. Once the beginner is comfortable imaging with an astro-video camera, then more advanced aspects of noise reduction associated with stacking can be explored using one of many excellent books or Internet sites on the subject.

M3 is an open cluster located in the Constellation Canes Venatici some 34,000 light years from Earth. It has a magnitude of 5.9 and an apparent diameter of 19 minutes (Fig. 7.7). The image of it below was made using *Deep Sky Stacker* and a median stack of 50 images. The camera settings were:

- Integration: X1024
- AGC: 0
- 3D-DNR: off
- Telescope: f/5 ST80A
- Mount: 4SE Equatorial Mode

Chapter 8

Outreach with Video Telescopes

Background

Some detractors describe video astronomy as "a solution looking for a problem." However, even they admit that video astronomy is excellent for outreach activities. Typically an outreach function will have a few telescopes set up for the public viewing. These telescopes are generally pointed at the Moon, Jupiter or Saturn, and maybe a deep space object.

Outreach activities often attract the public, and lines of people waiting for a few seconds of eyepiece time are not unknown. Eyepiece time constraints greatly restrict how many different objects can be viewed and the length of time someone can spend at the eyepiece. Add to the above focusing issues, hands moving telescope tubes off target, kids too short to reach the eyepiece, and the problems multiply. The event ends and people go home. Most are very happy having seen Jupiter, Saturn, and the Moon but also wondering what else is out there.

One single video telescope set up properly for an outreach function can reach hundreds of people simultaneously. Now, one can argue, with good justification, that such a mega event loses the personal touch that a one on one experience with a telescope provides. Also, events involving large numbers of people are beyond the capability of most astronomy clubs or sidewalk astronomers who want to reach out to educate the public. Professional observatories have the capacity for such events, but this book is aimed primarily at amateurs. Being so, we'll confine the discussion accordingly.

Video astronomy is not the cure-all nor is it appropriate for all occasions. However, it does have one aspect that greatly enhances an outreach event—two or more people can view through a telescope simultaneously. How many people depends upon:

- the objectives of the outreach event
- the constraints of the location where the outreach event is held
- the video telescope equipment available.

For an example, the output of one video telescope can be fed to a big screen or projection TV where a presentation is made to many people while the telescope slews from one object to another. On the other end, the video display could be a small 7- to 9-inch TV monitor located at the telescope that is viewable by only two or three people at one time.

Video Telescope Appropriateness

The objectives of an outreach effort define in several ways if and how a video telescope is appropriate and how it should be used. In deciding whether or not to use a video telescope in an outreach event consider some of the following:

- A video telescope can multiply viewing efficiency. This is important when significant numbers of people are expected to attend an outreach event and a limited number of telescopes are available.
- A video telescope eliminates the need for the public to touch the telescope to adjust focus or keep an object in view.
- A video telescope can isolate the public outreach effort from other ongoing activities, such as viewing and photographic activities by club members at a star party.

One aspect of video astronomy to keep in mind as an event is being planned is that TV sets and monitors emit light, lots of it. Also, the output of a video telescope is in color. Red screens are not a solution, especially if viewing nebulae are part of the planned activities. The TV monitor must be located in an area shielded from other ongoing astronomy-related activities, especially those that require the maintenance of night vision. Visitors should have access to and from the astro-video site that does not necessitate their passage through dark areas where astronomers are viewing or photographing the night sky with their telescopes.

So exactly what are some specific functions that a video telescope can do at an outreach event? Here are a few:

- allows a small number of people, say two to six people to view simultaneously through one telescope using a small TV or TV monitor.
- allows the telescope be located remotely from the TV or monitor
- allows a large group of people to view simultaneously on a large screen or projection TV.
- allows formal presentations.

Perhaps for most individual and small astronomy club outreach events, the most important attribute is the ability for a small group to view simultaneously through a telescope. The screen should be mounted close to the telescope but situated so there is no need for people to crowd around the telescope. A screen smaller than about 12 inches (30 cm) is difficult to see from a short distance away, and people will naturally tend to get close in order to see more. Increasing the size of the TV monitor helps, but the size and weight of larger TVs add to the logistics burden of stands/tables to set the TV on, electric power, etc. TVs no larger than about 22 inches are probably about as large as is practical for field applications.

Going in the other direction, a monitor smaller than 10 inches (25 cm) is difficult for more than three people to see at any one time, and one smaller than 7 inches (18 cm) is more suited for one person or two people if they stand very close to one another. The advantage of TV monitors of around 10 inches or less is they are easily battery powered, which is a big asset if someone is doing a street corner outreach where no power plugs are available.

Video Component Considerations

One aspect of video astronomy that many consider a negative is the need to use a GOTO mount. Very frequently 8- to 10-inch Dobsonian telescopes are used for outreach. These telescopes require minimal logistics support and are large enough to provide detailed views of objects in the night sky. Their biggest drawback is the need to manually keep the object in view of the eyepiece, something many members of the public have difficulty doing while they are viewing.

The attribute of a video telescope to increase a telescope's effective aperture has a major impact upon the size of telescopes needed for a public outreach activity. Small entry-level GOTO telescopes do very well with video. Telescopes such as the Meade ETX80, Celestron 102 SLT and 130 SLT, and the iOptron SmartStar R80A are ideal for viewing the brighter nebulae and star clusters, such as M8 and M45. In Europe, SkyWatcher has several small telescopes on its SynScan AZ GOTO mount that work very well. Several telescopes such as Orion's StarSeeker line, Meade's ETX 90, the Celestron's 90 SLT, and 127 SLT as well as its 4SE do very well for Solar System objects. Although lightweight mounts will have issues regarding imaging, they do very well for visual work.

For a major outreach event, a large screen or projection television can reach a large number of people simultaneously, especially if an auditorium or meeting room is available. However, an auditorium or meeting room is not necessary, as the TV monitor can easily be located outside. The meeting facilities do not need to be located at the same site or area as the telescope if Internet access is available at the telescope's location and at the meeting facilities.

An outreach can feature different themes, such as tonight's tour of deep space, nebulae in tonight's sky, a galaxy tour, a survey of different deep space objects, etc. This adds the need for a person to present the event who has both the knowledge and public speaking skills to keep the public's attention. A "mumbler" will kill the effort completely.

Although inexpensive entry-level telescopes are usable for such presentations, the accuracy of a German equatorial mount designed for astrophotography is a better approach. The public will soon lose attention if the mount or telescope operator is not accurate enough to do GOTOs that put objects in the field of view of the television display. Both the Celestron AVX and the iOptron ZEQ25 are excellent platforms for video astronomy.

Logistics

From an amateur's perspective regarding an outreach event, there are several issues that deserve consideration:

- event objectives
- public facilities
- crowd control
- public speaker
- electric power availability
- telescope operator skills
- stray light from TV monitors
- miscellaneous issues such as air traffic.

As stated earlier, the objectives of an outreach event will dictate the video telescope used. Even a street corner event can become more than just a look at the Moon, Saturn, and Jupiter. The video telescope can look into deep space even through the light pollution of a city. With an accurate GOTO mount, the street corner astronomer can include views of nebulae, open star clusters, globular star clusters, and interesting stars. The same is true for the local astronomy club. It, too, can expand the number and type of objects that it offers for viewing to the public. Some larger clubs can even put on narrated shows featuring deep sky objects in the sky. The type of event governs to a large extent the video equipment and support equipment needed.

Once the objectives of an outreach are defined, take a good look at the facilities and grounds available in relation to the expected public participation. Of importance is the location of facilities (meeting room, toilets, video telescope farm, video telescope TV monitor locations, etc.) as well as utilities, if any, that are needed for the event (Internet connections, electric power, etc.). Keep in mind that the video telescope, even if a remote monitor is used elsewhere, may have a small TV monitor at the telescope to facilitate star alignments, focusing, etc. If other activities such as those typical of a star party are going on, the video telescopes need a location or screening so that the light they produce does not interfere. The general public also needs a way to enter, participate, and then leave the location without a noticeable impact on any dark sky activities. The leaving part can be significant. People viewing the video telescopes will not have any decent night vision. The last thing anyone needs is for someone to stumble in the dark and damage someone else's astronomy kit. Associated with leaving is the automobile problem. The general public most likely will use their automobile headlights when they depart.

A laser is sometimes used at outreach events to point out where in the sky an object is located. The advantages of a laser pointer are readily apparent, especially if a formal presentation is being made. Keep in mind that law enforcement officials typically frown upon shining lasers into the night sky when aircraft are around and that pointing at an aircraft with a laser is illegal in most countries. If a laser pointer is used, a good idea is to designate one person as an aircraft spotter. The person doing the presentation can easily become too engrossed in the presentation to notice an approaching aircraft.

Another aspect to consider is the skill level of the people operating the video telescopes. Do they have the skills needed to accurately set up and align the telescopes? After the alignment process, can they rapidly slew from object to object without needing to fiddle around several minutes trying to put the object in the center of the field of view? Remember, the field of view with a video telescope is very small in comparison to an optical telescope.

Video Equipment Needed

What equipment is needed for a video telescope outreach event? The largest variables here are the telescope characteristics (aperture and focal length) followed by the type mount and video display. If the objects are very small and faint, such as many galaxies, then a telescope with a large aperture is needed to gather light as well as a moderate focal length of perhaps around 600–800 mm to produce an acceptable field of view. An accurate GOTO mount having the capacity to handle the weight of the telescope and sufficient GOTO accuracy to center objects with a relatively restricted field of view is without question needed.

A German equatorial GOTO mount with a payload capacity about twice the weight of the telescope and accessories is ideal, but a nice SCT on a heavy-duty dual tine fork mount with a wedge works very well too. The tripod must have sufficient weight capacity and stiffness to provide a stable platform for the mount and telescope. On the other hand, if the objects are some of the bigger and brighter deep space objects, such as most of the Messier and Caldwell objects, then a lightweight alt-azimuth GOTO mount and tripod with a small, fast, aperture telescope around 80–90 mm will work very well.

The next variable is the distance between the video telescope and the TV monitor displaying the image. An analog video signal can travel several hundred feet if a good cable is used. A USB digital signal is limited to much shorter distances. The exact distances are difficult to quantify. One train of thought is that if the distance between the camera and the TV display is more than 100–150 feet, then consideration should be given to broadcasting. This adds the complexity of an analog-to-digital converter, a computer or tablet at the site, a power supply for the computer, and a high speed Internet connection near the camera. However, most video telescope applications will have the TV monitor located close to the telescope, perhaps within 5–10 feet or so. A Windows 10 tablet with the latest Intel processor and USB interface can easily do a live video broadcast at the telescope that is available worldwide via the Internet, including a TV display several hundred feet away.

One astro-video camera manufacturer, MallinCam, offers a complete, portable outreach kit that, save for power sources, has all the required components excluding the telescope and mount. The kit costs about $2460 and contains the following components:

- MallinCam Jr PRO-EX camera (optional cooled camera is available)
- Focal reducer
- Camera cable
- WiFi AV transmitter (up to 150-m range)
- 21.5-inch broadcast standard TV monitor with flight case
- 110VAC to 12VDC power adapter (TV monitor)
- 12VDC cigarette lighter plug power adapter.

The kit requires no computer. The published range for the WiFi AV transmitter is up to 150 m (500 feet). This kit will also be excellent for home use when the astronomer wants to view deep space from inside the comforts of home. A cooled video camera is available as an option.

Urban Sidewalk Outreach Video Telescope

One outreach effort done by many amateurs is known as sidewalk astronomy or street corner outreach. These efforts generally involve a solitary amateur astronomer who sets up a telescope in a public area and invites people who pass by to step up and have a view at the stars. The nature of the event requires portable equipment.

One popular telescope for a sidewalk outreach event is an 8-inch Dobsonian; especially if the sidewalk location is walking distance from the astronomer's home. Small 5- to 6-inch aperture GOTO telescopes can easily be "backpacked," allowing the astronomer to reach the sidewalk location by walking, going on public transport, bicycling, or even driving with the equipment in the trunk of a car. However their limited aperture poses some issues viewing with the artificial sky glow associated with urban areas, and they are seldom used. There is also the issue of finding alignment stars within the severe artificial light domes associated with urban areas (see Appendix C in this book).

Inexpensive video equipment is a game changer for sidewalk astronomy. A video camera with a small 80-mm f/5 refractor on a GOTO alt-azimuth mount is very portable on foot, bicycle, public transport, etc. Small lithium battery backs can

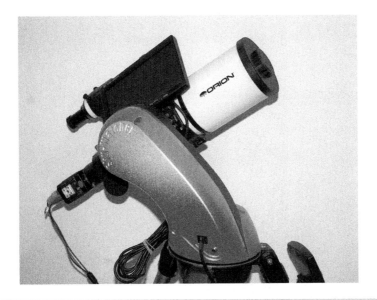

Fig. 8.1 Urban sidewalk outreach video telescope

easily power the video equipment as well as the telescope. The outreach kit can be as simple as a TV camera and a small 6- to 8-inch TV monitor or an 8-inch Win10 tablet, as shown in Fig. 8.1.

The video camera will easily display an image more than comparable that from an 8-inch telescope, especially in urban areas having a significant amount of artificial skyglow. Not only will the display be comparable to much larger telescopes, it will also be in color.

Chapter 9

Live Video Broadcasting

Equipment Needed

One interesting activity is going online and broadcasting the image in your telescope to people around our planet. The broadcast can be open to anyone who wishes to join or limited to the people you select. The actual set up and equipment needed is very basic. From the equipment perspective all that is needed is:

- an accurate GOTO mount and sturdy tripod
- a telescope
- an astro-video camera
- a camera cable
- an analog-to-digital converter (frame grabber)
- a computer (laptop or Windows10 Tablet)
- high speed Internet access.

In other words, all that is needed is the set up used for astro-video photography with the addition of access to the Internet. A computer or, as discussed in Chap. 10, a Windows 10 tablet is also required for the broadcast event.

Broadcasting Sites

There are two basic ways video astronomers broadcast on the Internet. Some set up their own private sites. This lets them control who watches and who does not. Others use sites specifically designed for amateur video astronomers. Here the site administrator controls who views and who does not. Which method is best is purely a personal preference.

For those who wish to set up their own private broadcasting site, an Internet search of "live video broadcasting" will provide many options and hosting sites to consider. Some of these sites offer free services, others charge a fee. For a private broadcasting site that will not cost you money, Google Hangouts on Air is reportedly both easy to set up and to use. Many of the Internet sites that host private broadcasting sites provide the capability to record a live show and then play it back on demand later.

Currently, three astronomy-related Internet sites exist that are online where amateur astronomers can do a live broadcast of what they are seeing in their video telescopes. None of these three sites charges amateurs any fees to use them:

- Video Astronomy Live: http://www.videoastronomylive.co.uk/
- Night Skies Networks: http://www.nightskiesnetwork.ca/
- AstronomyLive: http://astronomylive.com/.

Of the three sites, Video Astronomy Live has the most features, and given the resources of its commercial sponsor has the most polished interface. On the other hand, many people think that the Night Skies Network has the easiest interface to set up and use. It also has by far the greatest share of viewers and broadcasters. Astronomy Live is the newest site and is evolving. It is particularly useful for amateurs who already are using Google Hangouts on Air for broadcasting. One advantage of AstronomyLive is that the broadcast is recorded and can be played back on YouTube, etc. Which site is best? As with many things in life, the answer is personal preference.

Video Astronomy Live Website

"Video Astronomy Live" is a fairly recent site designed specifically for amateurs to broadcast live over the Internet what their video cameras are showing them. In addition to the live video, the site provides an on-line chat as well as an audio feed. Setting up a broadcast channel is simple and straightforward. The process used will take a couple of days at most, depending on how fast you respond and the time difference between your location and the administrator's location, which is located in the United Kingdom (http://www.videoastronomylive.co.uk/).

First you register on the site. Next you need to send an email to the site requesting that the administrator set up a broadcast channel for you on Video Astronomy Live. The site administrator will respond with a welcome email and a request that you provide the following information (note, you can provide this information in the first email you send to the administrator and shorten the time needed to set up a channel for you):

- *Channel Name:* What is the name you want to give to your channel? For example, the name of this author's channel is "The Yeller Dawg Observatory."
- *Introduction:* Introduce yourself to the public, who may want to watch your video show. Be as brief or long-winded as you care to be. You can look at the channels on the Video Astronomy Live website and see what others have written about themselves.

- *Your Location:* The name of the country and town, city, geographic area where your telescope is located. The Video Astronomy Live site will give the location of the site and also a weekly weather forecast for the area.
- *Images for the Image Gallery:* Attach any images of objects you have made with a video telescope that you may want to share. If you do not wish to show any images then skip this item.
- *Equipment List:* Identify the telescope, mount, focal reducers, camera, etc., you use for your astro-video kit.
- *Social Media Links:* List any social media links, if any, that you want to share with the public.

In addition to requesting this information, the site administrator will explain that you will need to use encoder software (all of which are free). More important, the administrator will send you your URL and your Stream Name. You will need to download the encoder software. Video Astronomy Live recommends one of the following:

- *Adobe Media Live Encoder* (FLME) Windows, Mac: http://www.adobe.com/ca/products/flash-media-encoder.html
- *XSplit Broadcaster* (XSplit) Windows: https://www.xsplit.com/
- *Open Broadcaster Software* (OBS) Windows, Linux, Mac: https://obsproject.com/.

Any one of the above will do. Of the three, *XSplit* is the most intuitive to use and is on par with *OBS* regarding features. FLME is the most constrictive and also the least intuitive. *OBS* is the only one of the three to support Linux and *XSplit* does not support Mac operating systems. Other encoder software is available. Most of the other software probably will also work with the Video Astronomy Live site. Figure 9.1 is a screen capture from *XSplit Broadcaster* during a broadcast.

Fig. 9.1 *XSplit Broadcaster* screen capture

Setting up the encoder software, at least for *XSplit* and *OBS,* is straightforward. Once you enter your URL and a Stream Name provided to you by the Video Astronomy Alive site administrator, you are ready to broadcast. To broadcast, simply run the encoder software you are using. To chat with viewers go to the Video Astronomy Live site, click on the Channels button and then on your site. You will now be able to communicate with viewers as well as monitor all that is happening with your broadcast.

Night Skies Network Website

The oldest and most widely known and used amateur astronomy broadcasting site is Night Skies Network: http://nightskiesnetwork.ca/. Live chat as well as audio are available. Anyone can log onto the site as a guest and view the broadcasts, if that is all they wish to do. Registering for viewing only is also an option and required if you wish to participate in the site's chat room, a place where you can chat real time to others about video astronomy-related topics.

However, in order to broadcast on the site you will need to register as a broadcaster and setup your system to meet Night Skies Network's criteria. The setup needed to be able to broadcast on Night Skies Network is straightforward and easy to do.

The very first thing is to register as a broadcaster. You will be asked to provide the following information:

- Channel name (the name of your observatory, for example)
- Password (the password you want to use for broadcasting and access to the Night Skies Network site)
- Your name
- A valid email address
- City (the location of your viewing site or observatory)
- Country (the country you are located in)
- Telescope type (what telescope(s) you are using for video astronomy.)
- Camera (what video camera you are using)
- Certification that you have read and understand Night Sky Network's rules.

Note Night Sky Network enforces their rules, especially the one prohibiting video captures requiring more than 3 minutes.

Night Sky Network will typically take 24–48 hours to process and either approve or reject your registration. After being notified that your registration is approved, you must set up your computer for broadcasting. Here's what you do:

1. The very first thing is to close all imaging software that you may have running on your computer, including programs such as AMCAP.
2. Connect your video capture device (frame grabber) to your computer. Needless to say, you need all the drivers for your camera and video capture device loaded onto your computer.

3. Log in to your channel using the user name and password that you selected in the application process. This will connect you to the site. A small window will pop up. The window will ask the question "Allow Nightskies network to access your camera and microphone?" Check the Allow and Remember boxes.
4. Another small window will pop up entitled Channel Chat. It will welcome you, tell you the status of your connection (extremely good, good, OK, or poor) including the latency.
5. An image should now appear on your screen. If not, right click on the view screen. A window will pop up giving you several options; choose settings.
6. Another window will now pop up. Click on the camera icon. This will give you a list of all the capture and grabber devices connected to your computer. Select the one you wish to use. Also adjust the microphone settings. (Sound can be muted but not deactivated.)
7. Left click on the view screen. A window will pop up that lets you adjust your image quality. Once you are satisfied with the image quality, click on the big green checkmark and then enter any equipment information you wish to show. When finished with your broadcast, click on the red button beneath the pop up window.

Astronomy Live Website

Astronomy Live is the newest and least known of the three astronomy video broadcasting sites (http://astronomylive.com/). It approaches broadcasting from an approach much different that the other two sites previously discussed. The site interface is easy to use once you understand how it works. Once you have signed up as a broadcaster you can arrange to broadcast over the Astronomy Live site. To do so, you have to use "Google Hangouts on Air," and to do this, you must have a Google account.

Before the broadcast is scheduled there are some things that must be done. First, go to the "Google Hangouts on Air" (https://plus.google.com/hangouts/onair) and create a "Hangout on Air." The hangout can be either immediate or scheduled for the future.

Next, complete the following form:

- Broadcast title
- Image of subject object (optional)
- Description of broadcast
- The link to the Google+ page for your Google Hangout on Air
- The link to your YouTube page for your broadcast.

In the "Google Hangouts on Air" link, provide the link to the page with your Hangout.

When you are ready to broadcast, visit the Google+ event page for your broadcast (https://plus.google.com/events). Click the "Start" button and tick off the box that you agree to the terms and conditions.

In the bottom right hand corner of the hangout is a button titled "Links." Click the button and then copy the embed code.

On the Astronomy Live page for your broadcast, click "Start broadcast," paste the embed code, and then save. Start the broadcast from your Google Hangout.

Planning and Executing a Live Broadcast

Executing a live broadcast can be a fun thing to do, especially if you get several viewers who join in on the ongoing conversation. The trick is to attract and keep your broadcast audience. This is where a carefully planned broadcast will help.

The first question to answer is, what is the subject of the broadcast? The possible subjects are quite numerous. One approach is to have a theme for the broadcast. A theme will focus the broadcast as well as provide a roadmap to follow during the night. A few suggestions for themes are:

- summer nebulae in the southern sky
- galaxies for the winter
- globular clusters
- tonight's tour of deep space
- a lunar trip.

Plan to spend 15 minutes or so at each object and encourage viewers to chat about the object and the astro-video kit used for the broadcast images. Take a look at the equipment you will use for the broadcast and the objects that are on your list to view. Make sure that the two are compatible. Doing a tour of the Moon with a short tube 80-mm refractor operating at f/2.0 certainly will not stir up interest like a tour of the Moon with 12-inch SCT at f/10 will.

Next, look at the weather forecast and your personal calendar. Decide what night you will do your show, providing the weather cooperates. Now make up your detailed viewing plan. A program such as *Stellarium* is a great tool for planning. Look at your viewing site and the location of the objects that you want to view. Make sure you don't schedule an object for viewing when some tree or structure may block it from view. As a general rule, arrange the order that you will view objects to minimize the time spent slewing between objects. This will shorten dead time waiting on telescope slews. Occasionally, other considerations may override and dictate a different approach.

Throwing a party where no one comes is not something most people want. A couple of days before the broadcast date start to check the weather and, if clear, start advertising your upcoming event. You have several resources available on the Internet, such as "the Astronomy Forum," the "Star Gazers Lounge," "Cloudy Nights," "Video Astronomy Forum," as well as the YAHOO video astronomy group "AstroVideoSystems."

In addition to these resources, the three video astronomy broadcasting sites also have an announcement feature where you can advertise your upcoming broadcast. All are excellent places to post an announcement a few days before your scheduled event. However, before posting an announcement on any forum, take the time to become a member of the forum as well as participate from time to time. Becoming known on the forums will also help increase viewership of your broadcasts.

The Rules

As the broadcaster, you also have the obligation to maintain order and decorum during your event. One essential role the person staging a live broadcast on the Internet has is that of the traffic cop. The astronomy broadcasting sites have rules and expect the broadcaster to abide by them, including keeping the people chatting in line. Keep in mind that most likely more people are viewing than have signed in as a viewer and that many of these people may be children.

Some people, mostly guys, seem not to be able to express themselves without the use of profanity. The users of such language should be warned to stop. If they persist, then use the site's resources to encourage them to quit. The NightSkies Network reportedly has a button you can push that will ban the offending viewer immediately.

However, the biggest offense that generally occurs during chats is an "off topic" conversation. This generally involves two viewers who accidently stray off topic and start having a conversation about last night's football game, politics, religion, some astronomy topic not even remotely connected to video astronomy, etc. As the broadcaster, you are expected to reel in these type conversations as soon as someone starts straying.

Generally all that is needed is a friendly reminder of "the rules."

Speaking of rules, the broadcast sites have published rules of behavior. These rules are very similar between the sites and mostly say use your common sense. Here is a synopsis of the rules typically used by astronomy broadcasting sites:

- Prohibiting the broadcasting of copyright materials is not only a rule typical among broadcast sites, but it is also the law. With video astronomy the largest offensive area regarding copyright materials is music. Although not intentional, often the broadcaster or even a viewer plays music that gets picked up by the broadcast. Also, some sites allow posting of photographs. Here you will find a considerable number of people who post images of others and totally disregard the intellectual property rights of others.
- The video astronomy broadcasting sites are for amateur astronomers to use. Commercial ventures are prohibited. Vendors have a lot of sneaky ways to infiltrate amateur astronomy sites. One typical strategy followed by some rather large and well known companies is sleeper members—employees who covertly join forums and who continuously recommend products made or sold by their company or employer.
- As mentioned earlier, discussions containing profane language, politics, or religion are not allowed. There are plenty of sites on the Internet where people having such interests can go.
- Finally, the amateur video astronomy sites do not allow unattended live streaming. This does not mean you cannot leave for a few minutes to take care of some biological function, get a hot coffee from the kitchen, etc.

NightSkies Network has a rule governing the kind of camera that can be used. The total integration time is limited to 3 minutes. The objective of the rule is basically to keep the people doing traditional astrophotography from using the site as a place to show off their work. The site is for near real-time viewing. It also requires the identification of equipment used to acquire and broadcast the image (telescope, mount, filters, camera, etc.).

Chapter 10

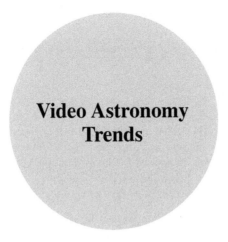

Video Astronomy Trends

The Tricky Business of Prediction

Trying to predict the future for a technology is a rocky road to travel. Technology and the public often have a habit of traveling down paths seldom dreamed of by people who should know. The introduction of $100 integrating video starlight cameras accelerated the expansion of amateur video astronomy. As newcomers to video astronomy explore uses for their cameras, who really knows what will turn up.

While visiting the Internet's astronomy forums one sees a marked increase in traffic regarding video astronomy. Most of the traffic deals with "how do I do it" questions asked by newcomers to video astronomy. Occasionally phrases such as 3D Astronomy pop up and disappear. Another technology application popping up on the Internet recently is the idea of using small, portable Windows10 tablets to eliminate laptop computers from the photographic process, whether it be traditional astrophotography or video astronomy. This seems a natural progression and similar to the way laptops, as their capabilities increased, banished large personal computers from the night sky not that many years ago.

One curious aspect of video astronomy is its analog nature in today's digital world. One can only wonder how much longer composite video will survive. Currently, CMOS camera technology is making inroads into the CCD camera world. Many believe that only in a few years CMOS cameras and their digital output will replace the current astro-video analog cameras. This trend seems established and well underway.

Finally the question is, "Will amateur astronomy as we know it today survive the urbanization of the world's population? By 2050 70% of the world's population will be in urban areas. Recent trends of replacing existing street lighting with high efficiency LED lamps shows a disturbing trend of increases in artificial sky glow over urban areas. How can the amateur astronomy community co-exist in the future world?

Three-Dimensional Astronomy

Current Status

Is three-dimensional (3D) astronomy something that will come to pass? It's probably far too early to answer that question. People who have experienced 3D astronomy seem positively impressed by the technology. Yet, 3D astronomy seems to remain in the incubator just waiting for its time.

Currently, four basic approaches to 3D astronomy are ongoing:

- Converting 2D photographic images into 3D images, then viewing them.
- Converting 2D video to 3D video and then viewing on a TV or computer screen.
- Converting live 2D video to 3D video and viewing near real-time on either a 2D or 3D TV.
- Converting optical telescopes and binoculars to provide a real-time 3D image.

Two of these approaches are passive, and the other two provide a real-time viewing experience. Except for using optical telescopes to obtain a three-dimensional image, all are anaglyphs and require anaglyphic eyeglasses to view in 3D. Three dimensional photographs and videos can be broadcast to a remote viewer over the Internet as live feed for near real-time video or as image files for typical video and photographic files.

Anaglyph describes the 3D effect obtained by using two different color filters to produce two different color images of the same scene at the same time. When viewed with one of the different color filters in front of each eye, a three-dimensional effect results. Red and cyan filters are typically used with red for the left eye. For convenience, these filters are placed in eyeglass frames, and these "anaglyph" eyeglasses are worn to see the three-dimensional effect.

Converting Two Dimensional Photographs into Three-Dimensional Anaglyphs

Several software packages can convert a typical 2D photograph into an anaglyph (a 3D image viewed with anaglyph glasses). In addition to programs like *Photoshop,* freeware programs are also available. Two such programs are:

- *Free 3D Photo Maker* by DVDVideoSoft Limited (http://dvdvideosoft-limited1.software.informer.com/)
- *Anaglyph Workshop* bySandy Knoll Software, LLC (http://sandy-knoll-software-llc.software.informer.com/)

Converting two-dimensional photographs into three-dimensional images requires no technology advances. The process is well developed and requires only a suitable computer program and the knowledge and skills needed to use it. As with other anaglyph applications, anaglyph eyeglasses are required to see the image in three dimensions. The final three-dimensional images can be viewed as paper photographs, on a computer monitor, or on a television screen.

Converting Two Dimensional Video into Three-Dimensional Anaglyphs

A variation of converting two-dimensional photographs into three-dimensional ones is converting two-dimensional videos into three-dimensional videos. This activity is far more popular than converting two-dimensional photographs. As with the conversion of two-dimensional photographs, the technology required to convert two-dimensional video files into three-dimensional video files is well developed and requires no technology advances.

An Internet search of "software convert 2D video into 3D video" will produce pages of hits. The first four pages will contain web pages identifying available software as well as sites providing primers and information about the process involved. Both commercial and freeware programs are available.

The primary driver behind this technology is converting home videos into three-dimensional videos for viewing on 3D television sets. Unlike the two- to three-dimension conversion software for photographs, video conversion software is typically simple to use with many programs and is fully automated. All that is required is to load the file in the program, and the program does the rest. The actual conversion can take some time, depending on the length of the video file and the speed of the computer. The memory requirements can be substantial for long videos. The final three-dimensional images can be viewed on a computer monitor or a television screen. As with the photograph conversion, the three-dimensional video programs are anaglyphs and require anaglyph eyeglasses to see the three-dimensional effect.

Near Real-Time Three-Dimensional Video Astronomy

Converting existing 2D photographs and video files into 3D can produce some nice photographs or videos to view from time to time. However, video astronomy is really about viewing the night sky live, in near real-time.

Technology developed primarily for the gaming industry allows the real-time conversion of two-dimensional computer games into three-dimensional games. This conversion is anaglyphic and requires anaglyphic eyeglasses to see in three dimensions. Some video astronomers are experimenting with this game technology to view galaxies, nebulae, and clusters in three dimensions and in real time. For want of a better description let's use the term "3D video astronomy" to describe a near real-time three-dimensional live video of deep space as well as planetary objects.

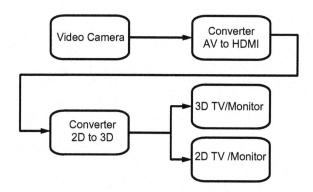

Fig. 10.1 Near real-time 3D component connections

To convert a video telescope into a 3D video telescope requires converting the analog composite signal from the astro-video camera into the HDMI digital format, then feeding the signal into a 2D to 3D converter that converts, real-time, the two-dimensional signal into a three-dimensional signal. Once converted, the digital signal is fed directly to either a conventional two-dimensional or to a three-dimensional TV that has a HDMI input or into a computer with an HDMI input port. Both, the analog to HDMI converter and the 2D to 3D converter are small devices that are a little bigger than the frame grabbers used for video astronomy. Anaglyphic eyeglasses are required to view objects in three dimensions (see Fig. 10.1).

The first step is to convert the analog composite video signal from the video camera into an HDMI digital signal. This converter is a piece of hardware similar to the frame grabber used for video astronomy; however, the frame grabber is not suitable for this task.

If you want to make your viewing location remotely from your telescope run an analog cable from the camera to the location of the computer, TV set, or TV monitor and do the conversion to an HDMI signal there. Which analog to HDMI converter should be used? At this time, not enough data are available to warrant recommending one particular converter over another. Many are available on the Internet. Here are some suggested specifications:

Description:

- AV/CVBS to HDMI converter
- Composite analog input AV/CVBS
- HDMI output 720p or 1080p
- Support PAL/NTSC video standards.

There are two kinds of HDMI converters available. Make sure that you purchase the type that converts an analog composite AV/CVBS (RCA) signal into a digital HDMI signal. The other kind does the reverse. It converts a digital HDMI signal into an analog composite (RCA) signal.

The output from the analog to HDMI converter is fed into a 2D to 3D converter that has an HDMI output. This piece of hardware is best for viewing the converted 2D video feed on a 3D TV using the glasses that came with the 3D television set. If a 3D television is not available then a 2D television/monitor can be used. The 2D television/monitor should have a resolution of 720p or 1080p to get a 3D effect. It must also have a HDMI input signal. Finally, the anaglyph eyeglasses that are included with the 2D to 3D converter must be worn to see the 3D effect when viewing with a 2D television set or monitor. Note the difference in the anaglyph eyeglasses. If you are using a 3D TV set, use the eyeglasses that came with the television set. If you are using a 2D television set, use the eyeglasses that came with the 2D to 3D converter.

Again, not enough data is available to recommend any particular 2D to 3D converter. However here are some attributes that deserve consideration:

- Input: HDMI
- Output: HDMI
- Resolution: up to 1080p
- 2D to 3D SBS-H (for viewing on a 3D television using the glasses provided with the television set)
- 2D to 3D amber/blue 3DRC (for viewing on a 2D television or monitor) using amber/blue eyeglasses
- Automatic 3D bypass (allows for the direct passage of a 3D signal to a 3D television)

Another consideration is what size TV monitor is best. If you are viewing where no power, other than batteries, is available, then a small HD TV monitor with an HDMI input and 720p or 1020P resolution will suffice. A 9- or 10-inch portable 12 vdc monitor will work, but some amateurs report that 13-inch or larger monitors produce the best 3D effect.

What is a rough cost estimate for the components required to assemble a 3D video astronomy system? Assuming that you already have a working astro-video telescope in operation and all you need is an analog to HDMI converter, 2D to 3D converter, and a monitor along with cables and a portable power supply, the cost is approximately $150 to $200 plus the cost of the display used (2D TV set/monitor or a 3D television set).

Real-Time Three-Dimensional Astronomy

Most amateurs do not use a video telescope to view the night sky. Instead they use the traditional optical system, a telescope with eyepieces. However, this does not mean that they, too, cannot see the night sky in three dimensions. Enter into the picture Denkmeier Optical, Inc., who produces quality eyepieces, binoviewers, diagonals, filters, and more for the amateur astronomy market.

Recently, Denkmeier Optical introduced its L-O-A 21 (Lederman-Optical-Array) three-dimensional eyepiece set on the market. The set consists of two eyepieces that are used with a binoviewer and a telescope or with binoculars that can use 1.25-inch telescope eyepieces. The eyepieces have a focal length of 21 mm and a field of view of 65 degrees with 14-mm eye relief. Looking toward the future, Denkmeier also has plans to manufacture three-dimensional binoculars.

The eyepieces produce a three-dimensional effect that has galaxies, nebulae, and star clusters floating in space. This is very similar to the effect described for three-dimensional video astronomy. The eyepieces are adjustable, allowing images in the center of the field of view to drift toward or away from the viewer. The three-dimensional effect is dependent upon a dark sky; the darker the sky, the more pronounced the effect. The eyepieces receive excellent reviews from both astronomy magazines and experienced amateurs alike.

However, do not expect to see the technology sweeping amateur astronomy. The eyepiece set costs $600 and then there is the price of a binoviewer. Even with discounts, the price approaches $1000, putting them on the "wish list" for most amateurs.

3D Video Astronomy Wrap-Up

3D video astronomy seems to excite people, but few are experimenting with it. Why? One reason appears to be the personalities of today's video astronomers. Until the introduction of the inexpensive integrating camera with extreme low light capabilities, video astronomy was the home of experimenters and tinkerers interested in exploring new worlds. Once inexpensive video astronomy cameras became available on the commercial market a new kind of astronomer entered the scene, one seemingly more interested in using video cameras for astronomy, astrophotography in particular, rather than exploring new worlds.

If a 3D astro-video system were offered on the commercial market at a reasonable price many amateurs would probably purchase one. However, few have the inquisitive temperament needed to research and develop their own system. Given the current state of the art, temperament of video astronomers, and the lack of a significant commercial demand 3D video astronomy is most likely a technology that will remain unexplored for a while, waiting for its time.

Windows 10 Tablets and Video Astronomy

Should You Use a Tablet with a Video Telescope?

Although not necessary for near real-time viewing of the night sky, computers are used extensively by video astronomers. The tasks they perform include controlling GOTO telescopes, focusing aids, camera control, data storage, data processing,

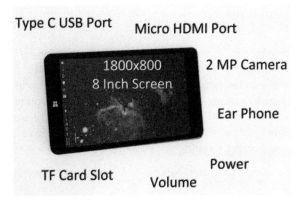

Fig. 10.2 Windows 10 tablet features

and broadcasting, among many. Using a computer in the convenience of your back yard, where electric power is an extension cord away as well as a table for the computer to rest on does not present any real logistical burdens. With video astronomy, the computer display can always displace a TV monitor with its power and support requirements, so the added burden of a laptop is not that bad. However, once one leaves the convenience of home, another set of rules associated with traveling light apply for many people, and some of these people are wondering why their small, convenient to use tablets cannot be useful in their hobby (see Fig. 10.2).

Tablets are rapidly increasing their computing capabilities coupled with decreases in prices. Many have Intel processors and can use the Microsoft Windows 10 operating system. Some even have dual operating systems (Android and Windows10). These Win10 tablets are capable of running the same Windows programs that run on more powerful laptop and desktop computers. This gives rise to the question, why not use a Windows 10 tablet instead of a laptop for video telescope applications where a computer is required?

Currently, when this question is asked on Internet forums, the reply is almost universally negative, with comments about processor speed, bandwidths, graphics cards, memory, and the like. However, mixed in with these comments are a few people who report success using astronomy-related windows programs with their tablets.

A casual look at Windows 10 tablets reveals a very broad range of sizes, prices, features, and computing capabilities. Unlike the iPad and Android world, the user base for Windows 10 tablets does not seem as enthused about their tablets. This lack of enthusiasm is evident by the very small number of "apps" developed for Windows tablets in comparison to the iPad and Android tablets. The primary attraction for Windows10 tablets appears to be "it runs the same Windows programs as do laptops and desktop computers." In the business world, this is a major attribute.

Current Tablet Usage

"It runs the same programs as do laptops and desktop computers" is an attribute that has caught the attention of some video astronomers. The traffic regarding using Windows 10 tablets on the astronomy forums indicates that a few amateur astronomers are beginning to explore using these small shirt pocket computers instead of a laptop computer in the field at night with some success. Here are some applications for Windows 10 tablets that amateur astronomers have posted on the astronomy-related Internet forums:

- video capture using a Celestron Neximage 5 camera then transfer to desktop for processing
- running the program *Stellarium* as a reference for objects in the sky
- running several programs simultaneously (no report of how the programs were used), such as *ASCOM, EQMOD, EQMOD-Game, The Sky/Cartes du Ciel, Stellarium with Stellarium Scope* or *SkyTools3 Professional, Nebulosity3, PHD2,* and *AstroTortilla*
- telescope control using a BlueTooth connection and *Stellarium, Stellarium Scope, ASCOM,* and *EQMod*
- image capture from a ZWO camera with *SharpCap* while viewing the images on a TV using the Microsoft Wireless Display Adapter
- processing video images using *PhotoShop* Elements
- mount, camera, and guider control with a HP Elitepad 1000

These applications are wide ranging. However, one issue with the above is that other than the HP Elitepad little information is provided about the capabilities of the tablets that were used.

Case Study of an Inexpensive Low-End Windows Tablet

A case study using an inexpensive tablet running Windows 10 was conducted to determine how well it could perform supporting a video telescope. The Chuwi Vi8 Plus was selected because it was the lowest priced Windows 10 tablet available at the time with an 8-inch screen and current state of the art technology. The tablet has a quad core CPU operating between 1.4 and 1.8 GHz. It has the latest type C USB port, which greatly improves data transfer speeds. Its graphics capability is good but, like most tablets, leaves gamers begging for more. The machine has 2 GB RAM, 32 GB ROM, and a 64 GB SD card. The Chuwi Vi8 is available over the Internet for $80 delivered from China to just about any location in the world.

The following computer programs were used to evaluate the suitability of the Chuwi Vi8 Plus to support astro-video applications:

- *XSplit Broadcaster*
- *Deep Sky Stacker*

- *GStar4Capture*
- *SharpCap 2.8*
- *Stellarium*

(*Note:* Time constraints prevented including *RegiStax 6* and *AutoStakkert!2*.)

Results

The Chuwi Vi8 tablet had no difficulty downloading and installing any of these programs. All but one of them ran at seemingly normal speeds without any glitches on the tablet. *Deep Sky Stacker* taxed the Chuwi Vi8 tablet. It ran but very slowly, about two minutes per light frame. The slowdown resulted when the program was forced to use virtual memory. The tablet had no difficulty with stacking using *SharpCap 2.8*. *GStar4Capture* and *XSplit Broadcaster* ran simultaneously with no conflicts. No difficulties were experienced using *XSplit Broadcaster* to broadcast live over the Internet.

The following non-astronomy programs also ran with no difficulty:

- *Microsoft PowerPoint 2016*
- *Microsoft Word 2016*
- *Microsoft Excel 2016*
- *Microsoft Edge 2016*

A tablet such as the Chuwi Vi8 PLUS is very capable of running current Windows programs, both astronomy-related and business-related. Whether or not its capabilities and features represent the minimum needed for video astronomy-related activities or they are overkill is an unknown at this time. However what is known is that the tablet is capable of running major astronomy-related programs used extensively by video astronomers.

If you are looking for a tablet to run astronomy programs in the field at night with your video kit, a Windows tablet meeting its specifications is a good starting point. In looking, consider that 2 GB of RAM is adequate, but 4 GB would be much better. Also 32 GB of ROM will be rapidly depleted if you save many video files, especially since half of the ROM is used by the Windows operating system and preloaded programs. Two Type C USB ports will also be helpful, as would be an HDMI input port.

It appears that even the lower-priced Win 10 tablets may be capable of controlling a telescope mount as well as capturing and then storing images. If true, this is a major capability for video astronomy. However before running out and buying a Win10 tablet to replace your laptop there are some things to consider:

- Unless you have a Type C hub to 3.0 USB-A OTG that supports both data transfer and tablet charging simultaneously, battery life is approximately only 3 hours.
- Windows programs take little advantage of touch screen technology. Using Windows programs without a mouse or touchpad is often tedious, especially when file management is required. (The Chuwi Vi8 tablet's Type C USB port can

handle a mouse and a frame grabber simultaneously at the cost of additional power consumption.)
- A small case for the tablet that also contains a Bluetooth keyboard is a very handy and inexpensive accessory.

Suggested Windows 10 Tablet Specifications

Specifications for a Windows tablet for video astronomy based upon the Chuwi Vi8 Plus are:

Overall Description

- Type: Tablet PC
- Operating System: Windows 10
- CPU: Intel Cherry Trail Z8300 Quad Core, 1.84 GHz
- GPU: Intel HD Graphic (Gen 8)
- RAM: 2 GB
- ROM: 32 GB
- SD card up to 64 GB
- Capacitive Screen (10-Point), IPS
- Screen: 8-inch. Resolution: 1280×800 (WXGA)

Ports:

- One SD card slot
- One Type C USB port
- One micro HDMI output port
- One 3.5-mm headphone jack
- Wireless
- WiFi: WiFi802.11b/g/n wireless Internet
- Bluetooth

Miscellaneous

- Dual 2.0 MP cameras (one front one back)
- Picture format: PNG, JPG, JPEG, GIF, BMP
- Music format: WMA, M4A, AAC, MP3, WAV
- Video format: 1080P, MP4
- Software: *MS Office: Excel, PPT, Word*
- E-book format:TXT

Conclusions

All in all the performance of the Windows 10 tablet used in the case study was very comparable to that provided by a laptop having a 2.2 GHz, dual core, two processor CPU. The major limitations of the tablet as tested were the lack of a keyboard and limited ports for peripheral devices. Here are the important points to remember:

- The Chuwi Vi8 PLUS and similar tablets are capable of running most Windows 10 programs, including the popular freeware programs associated with video astronomy.
- A Windows 10 tablet can perform most of the functions done by larger laptops.
- 2 GB RAM is adequate.
- 32 GB ROM is marginal, as the operating system and other programs take over 15 GB of ROM.
- An 8-inch screen is rather small for reading program menus, etc. A 10-inch screen is a better option.
- The Type C port can both power and pass live video data input from a frame grabber with no noticeable lag.
- The tablet is capable of live video broadcasts.
- The tablet can support live stacking.

The tablet's suitability for video astronomy would be greatly enhanced if it had 4 GB of RAM, 64 GB of ROM, 2 Type C USB ports, and either a two-way HDMI port or a HDMI input port and a HDMI output port. In any case, as configured the tablet can capture, stack, store, broadcast, and display images on its screen. For the price typically charged for an 8-inch TV monitor used for video astronomy, you can have a Windows 10 Tablet that has a high definition 8-inch screen for viewing objects in the night sky as well as the computing capability required for imaging and broadcasting with a video telescope.

Will Windows tablets ever replace laptops? To a significant extent, yes, perhaps as much as 50%. Tablets are particularly suited for video astronomy usages that do not involve control of telescopes and other accessories.

Note The dual operating system (Windows/Android) tablets do not share computer resources. Both RAM and ROM memory is split evenly between the two operating systems. If the tablet has 4 GB RAM and 64 GB ROM, the resources are split giving 2 GB RAM and 32 GB ROM to each operating system.

Digital Astro-Video Cameras

One reasonable assumption is that video astronomy will migrate to the digital world. In fact the migration has already begun. From a commercial perspective, video astronomy will likely never have a market sufficiently large to entice the design and manufacture of sensors specifically for the amateur astronomy market.

This means that video astronomy will piggyback on technology developed for other applications, as it now does with low light CCTV cameras designed for the security industry. The trend for the CCTV industry is toward digital cameras. Also, a very significant percentage of recent newcomers to video astronomy seem more interested in using video astronomy for low-cost imaging rather than for the simple pleasure of observing the night sky. For them, analog video devices cannot provide the color depth, definition, and resolution associated with modern high-definition digital devices.

MallinCam recently introduced an interesting, all-digital astro-video starter's kit called the SkyRaider Netbook Kit. For $300 they offer a digital video-astronomy starter kit complete with camera and a 10-inch Windows10 netbook computer with all cables needed. The kit is ready to go out of the box, with the computer's battery charged and all programs installed. Another camera manufacturer, Astro Video Systems, recently announced a new video camera that will have both HDMI and XVGA outputs. These are only two of several digital-related video camera technologies currently available. Within the traditional astrophotography camera community, several existing manufacturers of astronomical CCD cameras have recently introduced digital, low-light video cameras that compete very well with the current analog astro-video cameras

Current trends point toward the likely future of video astronomy being essentially an all-digital world. Current digital devices do not have the transmission distances using cables that are associated with analog cameras. However, wireless digital technology can provide the means to transmit signals 100 feet (30 m) or more. Substitute a Windows 10 tablet or a netbook for the laptop and the following system is very likely:

- Digital video camera
- WiFi transmitter
- Windows 10 tablet/netbook

Urban Astronomy

As stated earlier, recent studies indicate that the worldwide trend toward energy-efficient LED street and security lighting may actually increase the levels of light pollution associated with urban areas. Unlike sodium- and mercury-based lighting technologies that emit light in fairly narrow bandwidths that can be filtered, light from LED fixtures is broadband. Apparently many LED lighting projects worldwide are essentially bulb replacements with little thought given to color, fixture design, fixture placement, and other aspects required for a good lighting design. The result is that lighting levels on roadways increase and fixture placements direct light upward, which helps no one.

From an urban astronomy perspective trespass light has increased as well as sky glow, both generated by a broadband light that cannot be attenuated by filtering. How this will impact video astronomy as well as astronomy in general is uncertain. At this time, insufficient data exists to quantify the effects upon video cameras.

Fig. 10.3 M31, the Andromeda Galaxy

Theoretically video cameras should perform similar to the example discussed in Chap. 4 for broadband unfiltered artificial sky glow. Although the video camera cannot eliminate the impact of artificial sky glow, it can offer some mitigation and allow astronomers some access to deep space. This capability of a video camera to detect objects buried in artificial sky glow is an incentive for urban amateurs to switch to video astronomy. The unanswered question remaining is, will the views produced by video telescopes be sufficient to satisfy a significant percentage of the amateur astronomy community as a whole? If so, the future of video astronomy looks bright. In any case, the majesty of objects such as the Andromeda Galaxy will be lost to many astronomers (see Fig. 10.3).

Note Further information related to light pollution is available from the International Dark Sky Association (http://darksky.org/about/).

Appendix A

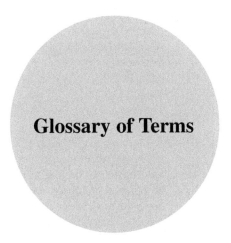

Glossary of Terms

3D-DNR (three-dimensional digital noise reduction)	A process available with current low-light CCTV cameras to reduce noise through a combination of image stabilization, digital image enhancement, dark frame subtraction, and stacking.
Absorption nebulae	See Nebulae.
Alt-azimuth mount	This mount is aligned with the plane of Earth. One axis is pointing to a spot in space located directly above an observer called the zenith and the other axis is parallel to the plane of Earth. As with an equatorial mount, each object in the sky has its own set of coordinates, but these change with the location of the observer on Earth and the time of the day. Because an alt-azimuth mount is not aligned to the celestial pole, movements in both azimuth and altitude are required to keep an object centered in view of an eyepiece as Earth rotates. These adjustments constantly change as Earth rotates; thus, a computer is needed to calculate and control the motions of the mount to keep an object centered in an eyepiece or a camera.
Aperture	The diameter of a telescope's light-collecting surface—its primary mirror or lens—and defines a telescope's light-gathering ability as well as its ability to resolve details.

Astronomical unit (AU)	The average distance between Earth and the Sun, about 149.6 million km. The nearest star to our Sun, Alpha Centauri Proxima, is about 271,000 AU from Earth.
Astro-video camera	A small television camera used with a telescope instead of its eyepiece. It displays an image using either an NTSC or a PAL television standard analog signal.
Back focus	The distance from the end of the drawtube of a telescope to the telescope's focal plane.
Cassegrain telescope	A telescope that strongly resembles the Newtonian telescope except instead of a flat diagonal mirror to reflect light out the side of the telescope's tube, the design uses a convex mirror to reflect light back down the tube, where it passes through a hole in the center of the primary mirror to the focuser and eyepiece. The Cassegrain telescope was essentially ignored until after the mid 1900s when a method was developed to cheaply manufacture a variation called the Schmidt Cassegrain Telescope. Today, two versions of the Cassegrain telescope, the Schmidt Cassegrain and the Maksutov Cassegrain telescopes are in wide usage among amateur astronomers.
Celestial sphere	An imaginary sphere of a large but undetermined radius with Earth at its center that is aligned with Earth's poles and equator. To an observer on Earth, all celestial objects appear attached to the inside surface of the celestial sphere.
Chromatic aberration	The inability of a telescope to bring all wavelengths of light to a common focus.
Closed circuit television camera (CCTV)	A small inexpensive television camera typically used as a security camera. Low-light CCTV cameras, often called starlight cameras, are used as video cameras by amateur astronomers. They have excellent sensitivities with a minimum illumination as low as 0.0001 Lux for monochrome and 0.001 Lux for color output. Unlike Digital Single Lens Reflex (DSLR) cameras and astronomical CCD cameras typically used for astrophotography that have digital outputs, a CCTV camera has an analog output signal that complies to either the NTSC (*National Television System Committee*) or PAL (*Phase Alternating Line*) television standard. Which standard is used is dependent on the geographic location. The effective frame (complete image)

Appendix A: Glossary of Terms 157

	refresh rate for an NTSC camera is 30 frames per second and 25 frames per second for a PAL camera using a process called interlacing. Either frame rate is sufficient for our brains to trick us into seeing both images as one on a TV set—in other words, a complete image.
Constellations	The International Astronomical Union (IAU) divided the celestial sphere into 88 regions called constellations. The borders of some constellations are arbitrary, while the borders of others follow patterns in the sky often related to ancient myths.
Deep space objects	Any celestial object not in orbit around the Sun; typically called a DSO by astronomers. Here are the main categories of deep space objects: stars, open clusters, globular clusters, galaxies, and nebulae.
Emission nebulae	See Nebulae.
Equatorial telescope mount	A mount aligned with the celestial sphere. This allows the mount to track an object as it crosses the night sky while keeping the object centered and stationary in an eyepiece or in a camera. Since the equatorial mount is aligned with the celestial pole, the rotation of objects in the sky is at a constant rate—15 degrees per hour. Each object in the sky has its own unique set of celestial coordinates that remain constant regardless of the time of day or location of the observer on Earth. If the equatorial mount is precisely aligned with the celestial pole, a motorized mount can keep an object stationary long enough for exposures exceeding several minutes or more with no guidance. Since the movement is constant and in one direction only, no computer is needed to track an object as Earth rotates on its axis. Two major equatorial mount variants are used in amateur astronomy: an alt-azimuth mount on a wedge that aligns the mount with the celestial pole or a German equatorial mount (GEM), which is designed with its axes of rotation aligned with the celestial sphere.
Field of view (FOV)	The area of the sky covered by an image. This area is described in arc-minutes in diameter if it is circular or in arc-minutes of length and width if it is rectangular. Since an arc-minute is a known unit of measurement, it provides a way to accurately compare the variants in field of view caused by different camera senor sizes and telescope focal lengths. The Moon is often used as a reference point, as it covers about 30 arc-minutes of the night sky.

Field rotation	Since an alt-azimuth mount is not aligned with the celestial sphere, objects will rotate in the eyepiece or camera as Earth rotates. This rotational effect is called field rotation. Although too slow to be noticeable for visual work, it has two major impacts upon alt-azimuth mount photography. It limits the duration of exposures that can be made and also limits the areas of the night sky where objects to be photographed can be located. For visual astronomy field rotation is not an issue. However, for photography field rotation results with stars making trails in the image and blurring the object being photographed. The rate of field rotation is dependent upon the latitude of the observer on Earth's surface, with a maximum for an observer located on Earth's equator and a minimum, zero, for an observer located at Earth's poles, where the zenith and celestial poles are one in the same. In addition, the rate of field rotation is also dependent on the location of an object in the night sky, with the maximum rate occurring at the zenith and the minimum rate toward the eastern and western horizons. Since the location of an object in the sky is constantly changing, the rate of field rotation associated with an object's position is also constantly changing.
Focal length	The distance between the center of a lens or curved mirror and its focus.
Focal ratio	The ratio between the focal length of a telescope and its aperture (focal length divided by the aperture). The focal length provides an indication of image size and field of view while the focal ratio provides information about image brightness.
Galaxy	The ultimate collection of stars ranging from hundreds of millions to trillions of stars. Galaxies come in various shapes, sizes, and ages. Astronomers place them into one of the following five categories: spiral, bar, elliptical, lenticular, and irregular.
Globular star cluster	Tightly packed balls of stars, ranging from tens of thousands to hundreds of thousands of stars. They tend to form a halo around galactic centers and contain very old stars approximately 10 billion years old.
GOTO mount	The term GOTO refers to a motorized, computer-controlled telescope or camera mount that can automatically find and then track objects in the night sky. Typically the computer used is a small handheld device, about the size of a small portable telephone,

Appendix A: Glossary of Terms

	with controls for using the mount. You tell the computer what you want to see and the computer will then "go to" the object and keep it centered in the eyepiece while you view. GOTO mounts can be either an alt-azimuth or equatorial.
Image size	The ratio between the telescope's focal length and the eyepiece's focal length or the diagonal measurement of a camera's sensor. For an example, a telescope with a focal length of 2000 mm will produce a magnification of 80 with a 25-mm eyepiece ($2000 \div 25 = 80$).
Light year (ly)	A light year is the distance that light travels in a vacuum over a period of one Earth year. Alpha Centauri Proxima is about 4.2 light years from Earth.
Messier objects	A Frenchman (1730–1817), Messier is perhaps one of the most widely known astronomers in history yet he never wrote a major book or scientific treatise. His career was finding and observing comets. In the course of doing so, he constantly found objects in the sky that looked like comets but later proved otherwise. To prevent others from mistaking these objects for comets, he compiled a list of the objects he observed as well as those observed by others. Unlike many other lists of that time, Messier was meticulous about accuracy and verification. The Messier list is still widely used today by amateurs worldwide. The currently accepted version of the list has 110 objects and consists of 40 galaxies, 29 globular clusters, 28 open clusters, 10 nebulae, and 3 other objects such as the Sagittarius Star Cloud.
Nebulae	Large clouds of gas or dust typically having dimensions measured in light-years. Often you will see the terms galactic nebula and diffuse nebula. A galactic nebula is one that is within our galaxy, the Milky Way. A diffuse nebula is extended with no well-defined boundaries, a definition that fits most nebulae. Nebulae fall into three basic type regarding their visibility. Emission nebulae are gas clouds that become visible when the gas molecules are excited by high energy photons from a nearby bright star causing them to emit visible light. Reflection nebulae are dust clouds that are visible by reflecting the light of a nearby bright star. Absorption nebulae

	are clouds of dust so thick that they obscure the light of background stars or nebulae.
New General Catalog (NGC)	Perhaps the most widely used night sky catalog this was compiled by John Dreyer in 1888. It contains 7840 deep space objects. Over the years Dreyer published two supplements to the NGC that he called Index Catalogs (IC). The first IC contained 1520 additional objects and the second IC added another 3866 objects for a total of 5386 IC objects.
Newtonian telescope	In 1668, Sir Isaac Newton devised a different way to view the stars. Instead of a glass lens to gather and concentrate light, he used a parabolic mirror to do the job. Light entered the telescope tube and was reflected off a parabolic mirror mounted at the end of the tube back up the tube to another smaller flat mirror mounted at a 45 degree angle and then out of the side of the tube through a focuser and eyepiece. This type of "reflecting" telescope is very popular among amateur astronomers. Large mirrors are easier to make and lighter than large lenses; thus, large and medium size reflecting telescopes are cheaper to manufacture or easier for an amateur astronomer to make than are refractors.
NTSC (National Television System Committee)	*See Closed* Circuit Television Camera.
Open star cluster	Stars in close proximity to one another that form irregular but often interesting patterns. No hard-fast rule exists regarding how many stars are in an open star cluster, and the number can vary widely. Open star clusters typically have mostly young stars. This is because as the eons pass, the stars in a cluster disperse, leaving few clues about the cluster or the stars it contained. Perhaps the most famous open cluster is M45, also known as the Pleiades or the Seven Sisters. This object is easily visible with the naked eye and was well known throughout recorded history.
OSD (on screen display)	OSD is a pull-down menu displayed by a CCTV camera on its monitor that allows the operator to adjust camera settings.
PAL (phase alternating line)	See Closed Circuit Television Camera.
Parsec (pc)	A parsec (pc) is the distance from Earth to a theoretical point that produces an annual parallax of one arc-second in the sky (about 3.26 light years). Alpha Centauri Proxima is about 1.3 parsecs from Earth.

Appendix A: Glossary of Terms

Payload	"Payload" and "rated payload" are terms often heard connected with telescope mounts. Rated payload is the amount of weight the manufacturer says that a mount can carry and still perform as specified. It does not include the mount's counterweights, the weight of the mount itself, or its tripod, polar scopes, etc. Payload is the total weight of the telescope, camera, adapters, focal reducer, finder, etc., that is carried by the mount. Although the rated payloads for equatorial mounts tend to be accurate for visual work, astrophotographers find they face fewer issues if they limit the weight they put on a mount to 50–60% of the mount's rated payload.
Published magnitude	The magnitude of a star cluster, nebula, or galaxy would have if it were shrunk to a point source like a star. The process used to express brightness for star clusters, nebulae, and galaxies is different than that for stars. Here it is the object's apparent total visual brightness expressed as a magnitude using the same brightness scale used for stars. However, stars are point sources, while star clusters, galaxies, and nebulae have an apparent area that is measurable from Earth, or else they would appear the same as stars to us. Some objects are quite large and others small.
Refractor telescope	The "spyglass" the pirates of old used to look for other ships. A refractor telescope has a long tube and has a large lens on the front end to gather and concentrate light and an eyepiece and focuser at the other end. This is the kind of telescope Galileo used 400 years ago when he gazed upon the heavens and changed forever how humanity viewed the universe.
Sky glow	The sky at night is never truly dark but has a faint glow that astronomers call "sky glow." Light to produce this glow comes from three basic sources. One obvious source is light from the stars. The second source is not so obvious. Oxygen in Earth's atmosphere is excited by photons from the Sun and has a greenish glow that can be photographed from outer space and, under certain conditions, from here on Earth. This glow is not the famous northern or southern lights (Aurora Borealis or Australis), as they are another phenomenon that is localized and is not included in basic sky glow calculations. However, when present, they cannot be ignored. The last source of light is artificial sky glow, the

	light pollution produced by humans from street lights, automobile head lights, advertising signs, security lights, etc.
Solar System objects	The Sun and those objects in space under the influence of its gravity. This includes the planets with their moons, asteroids, comets, and objects in the Kuiper Belt and the Oort Cloud as well as manmade objects in space.
Stacking	Aligning two or more images of an object, taken with the camera and object being relatively stationary, so that each pixel for each image is aligned with the identical pixels for the other images. In other words, the images are stacked on top of one another. The values of each pixel stack are then either added together in what is known as a linear stack and the total value of the stacked pixels is then used for the final image, or they are averaged in what is known as a statistical stack, and the average value of the stacked pixels is then used for the final image. (Some statistical routines use the median value of the stacked pixels.)
Stellar magnitude	Star brightness is measured using a scale called stellar magnitude. This is a measure of the apparent brightness of a star as viewed from Earth, not its actual brightness. A star having a magnitude of 0.0 is 2.512 times brighter than a star with a magnitude of 1.0; a 1.0 magnitude star is 2.512 times brighter than a star with a magnitude of 2.0; and so forth. The larger the number, the dimmer the star. The magnitude scale has negative numbers. Our Sun has a magnitude of -27, the full Moon has a magnitude of -13, and Sirius, the brightest star, has a magnitude of -1.4.
Surface brightness	Published magnitude presents a problem in that the surface areas of objects vary considerably. When the light is spread over the total visible surface area, the object is much dimmer than the published magnitude. In many cases an object with a large surface area and bright magnitude that may even indicate naked eye visibility is actually too dim to view in a small- or medium-sized telescope or vice versa, whereas an object with a small surface area and a dim magnitude may be easily seen in a small telescope. For visual observation and photography, surface brightness is more meaningful. Galaxies and nebulae are distributed light sources. Their brightness can be represented as a ratio of their total mag-

	nitude divided by their surface area. This ratio is called "surface brightness" and is expressed in magnitude per square arc-second.
Video astronomy	The substituting of the eyepiece of a telescope with an astro-video camera in order to live view the night sky.
Video telescope	The combination of a telescope with an astro-video camera. The camera provides a live, analog, video signal to a television monitor, DVD player, etc., that can be shared by more than one person. More complex configurations involving analog-to-digital converters, computers, internet access, etc., are also used, depending on exactly what the astronomer wants to accomplish.

Appendix B

Maximum Exposure Time Tables Based on 0.125 Degrees of Field Rotation

Maximum exposure time in seconds for an observer at 0° north or south latitude

Object's azimuth angle (degrees)			Object's altitude angle (degrees)								
			10	20	30	40	50	60	70	80	90
			Maximum exposure time in seconds								
0	180	360	30	28	26	23	19	15	10	5	0
10	170	190	350	30	29	26	23	20	15	10	5	0
20	160	200	340	31	30	28	24	21	16	11	6	0
30	150	210	330	34	33	30	27	22	17	12	6	0
40	140	220	320	39	37	34	30	25	20	13	7	0
50	130	230	310	46	44	40	36	30	23	16	8	0
60	120	240	300	59	56	52	46	39	30	21	10	0
70	110	250	290	86	82	76	67	56	44	30	15	0
80	100	260	280	170	162	150	132	111	86	59	30	0
90		270		339	323	298	264	221	172	118	60	0

Appendix B: Maximum Exposure Time Tables Based on 0.125 Degrees of Field Rotation

Maximum exposure time in seconds for an observer at 10° north or south latitude

Object's azimuth angle (degrees)				Object's altitude angle (degrees)							
				10	20	30	40	50	60	70	80	90
				Maximum exposure time in seconds							
0	180		360	30	29	26	23	20	15	10	5	0
10	170	190	350	30	29	27	24	20	15	11	5	0
20	160	200	340	32	30	28	25	21	16	11	6	0
30	150	210	330	35	33	30	27	23	18	12	6	0
40	140	220	320	39	37	34	30	26	20	14	7	0
50	130	230	310	47	45	41	36	30	24	16	8	0
60	120	240	300	60	57	53	47	39	30	21	11	0
70	110	250	290	88	84	77	68	57	45	30	15	0
80	100	260	280	173	165	152	134	113	88	60	30	0
90		270		344	328	303	268	225	175	120	61	0

Maximum exposure time in seconds for an observer at 20° north or south latitude

Object's azimuth angle (degrees)				Object's altitude angle (degrees)							
				10	20	30	40	50	60	70	80	90
				Maximum exposure time in seconds							
0	180		360	31	30	28	24	21	16	11	6	0
10	170	190	350	32	30	28	25	21	16	11	6	0
20	160	200	340	33	32	29	26	22	17	12	6	0
30	150	210	330	36	35	32	28	24	18	13	6	0
40	140	220	320	41	39	36	32	27	21	14	7	0
50	130	230	310	49	47	43	38	32	25	17	9	0
60	120	240	300	63	60	55	49	41	32	22	11	0
70	110	250	290	92	88	81	72	60	47	32	16	0
80	100	260	280	181	173	159	141	118	92	63	32	0
90		270		361	344	317	281	235	183	125	64	0

Maximum exposure time in seconds for an observer at 30° north or south latitude

Object's azimuth angle (degrees)				Object's altitude angle (degrees)							
				10	20	30	40	50	60	70	80	90
				Maximum exposure time in seconds							
0	180		360	34	33	30	27	22	17	12	6	0
10	170	190	350	35	33	30	27	23	18	12	6	0
20	160	200	340	36	35	32	28	24	18	13	6	0
30	150	210	330	39	38	35	31	26	20	14	7	0
40	140	220	320	45	42	39	35	29	23	15	8	0
50	130	230	310	53	51	47	41	35	27	18	9	0
60	120	240	300	68	65	60	53	45	35	24	12	0
70	110	250	290	100	95	88	78	65	51	35	18	0
80	100	260	280	196	187	173	153	128	100	68	35	0
90		270		391	373	344	304	255	199	136	69	0

Appendix B: Maximum Exposure Time Tables Based on 0.125 Degrees of Field Rotation   167

Maximum exposure time in seconds for an observer at 40° north or south latitude

Object's azimuth angle (degrees)			Object's altitude angle (degrees)								
			10	20	30	40	50	60	70	80	90
			Maximum exposure time in seconds								
0	180	360	39	37	34	30	25	20	13	7	0
10	170	190	350	39	37	34	30	26	20	14	7	0
20	160	200	340	41	39	36	32	27	21	14	7	0
30	150	210	330	45	42	39	35	29	23	15	8	0
40	140	220	320	50	48	44	39	33	26	17	9	0
50	130	230	310	60	57	53	47	39	30	21	11	0
60	120	240	300	77	74	68	60	50	39	27	14	0
70	110	250	290	113	108	99	88	74	57	39	20	0
80	100	260	280	222	212	195	173	145	113	77	39	0
90		270		442	422	389	344	289	225	154	78	0

Maximum exposure time in seconds for an observer at 50° north or south latitude

Object's azimuth angle (degrees)			Object's altitude angle (degrees)								
			10	20	30	40	50	60	70	80	90
			Maximum exposure time in seconds								
0	180	360	46	44	40	36	30	23	16	8	0
10	170	190	350	47	45	41	36	30	24	16	8	0
20	160	200	340	49	47	43	38	32	25	17	9	0
30	150	210	330	53	51	47	41	35	27	18	9	0
40	140	220	320	60	57	53	47	39	30	21	11	0
50	130	230	310	72	68	63	56	47	36	25	13	0
60	120	240	300	92	88	81	72	60	47	32	16	0
70	110	250	290	134	128	118	105	88	68	47	24	0
80	100	260	280	265	253	233	206	173	134	92	47	0
90		270		527	503	464	410	344	268	183	93	0

Maximum exposure time in seconds for an observer at 60° north or south latitude

Object's azimuth angle (degrees)			Object's altitude angle (degrees)								
			10	20	30	40	50	60	70	80	90
			Maximum exposure time in seconds								
0	180	360	59	56	52	46	39	30	21	10	0
10	170	190	350	60	57	53	47	39	30	21	11	0
20	160	200	340	63	60	55	49	41	32	22	11	0
30	150	210	330	68	65	60	53	45	35	24	12	0
40	140	220	320	77	74	68	60	50	39	27	14	0
50	130	230	310	92	88	81	72	60	47	32	16	0
60	120	240	300	118	113	104	92	77	60	41	21	0
70	110	250	290	173	165	152	134	113	88	60	30	0
80	100	260	280	340	325	299	265	222	173	118	60	0
90		270		678	647	596	527	442	344	235	120	0

Appendix B: Maximum Exposure Time Tables Based on 0.125 Degrees of Field Rotation

Maximum exposure time in seconds for an observer at 70° north or south latitude											
Object's azimuth angle (degrees)				Object's altitude angle (degrees)							
				10	20	30	40	50	60	70	80	90
				Maximum exposure time in seconds							
0	180		360	86	82	76	67	56	44	30	15	0
10	170	190	350	88	84	77	68	57	45	30	15	0
20	160	200	340	92	88	81	72	60	47	32	16	0
30	150	210	330	100	95	88	78	65	51	35	18	0
40	140	220	320	113	108	99	88	74	57	39	20	0
50	130	230	310	134	128	118	105	88	68	47	24	0
60	120	240	300	173	165	152	134	113	88	60	30	0
70	110	250	290	253	241	222	196	165	128	88	45	0
80	100	260	280	497	475	437	387	325	253	173	88	0
90		270		991	945	871	771	647	503	344	175	0

Appendix C

Star Charts for Urban Areas with Significant Light Pollution

Appendix C: Star Charts for Urban Areas with Significant Light Pollution

Northern Hemisphere at Latitude 40 Degrees

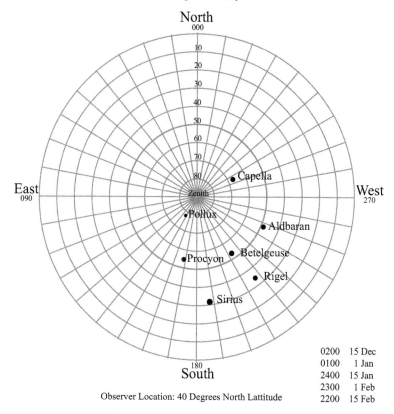

Fig. C.1 North 40 degrees January

Appendix C: Star Charts for Urban Areas with Significant Light Pollution

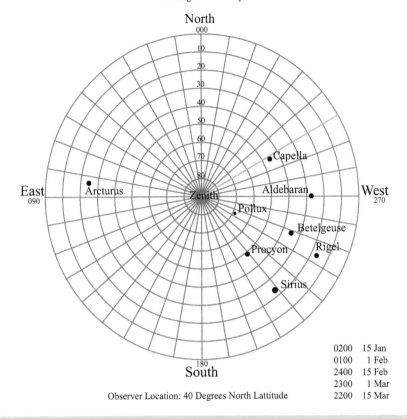

Fig. C.2 North 40 degrees February

172 Appendix C: Star Charts for Urban Areas with Significant Light Pollution

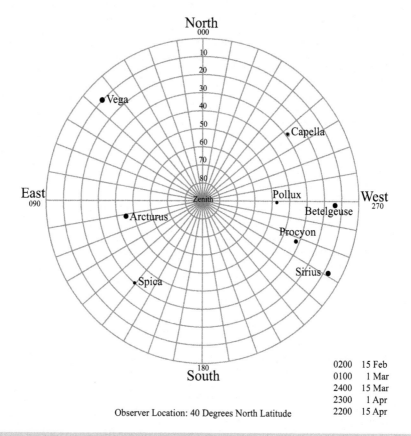

Fig. C.3 North 40 degrees March

Appendix C: Star Charts for Urban Areas with Significant Light Pollution 173

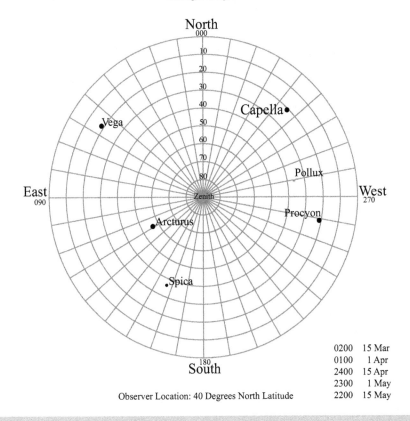

Fig. C.4 North 40 degrees April

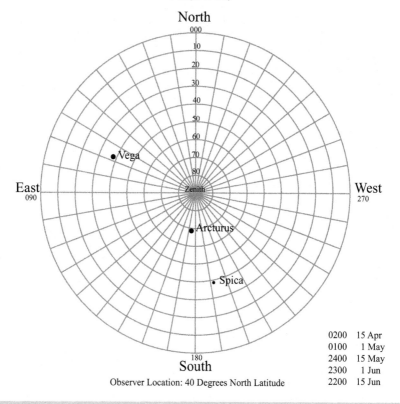

Fig. C.5 North 40 degrees May

Appendix C: Star Charts for Urban Areas with Significant Light Pollution

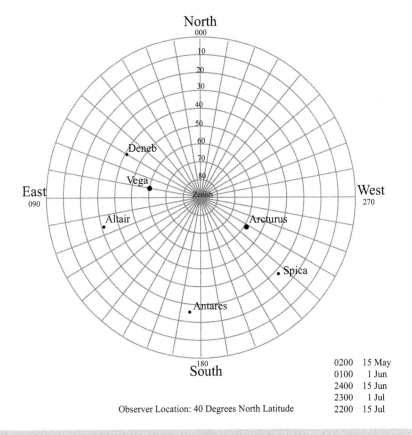

Fig. C.6 North 40 degrees June

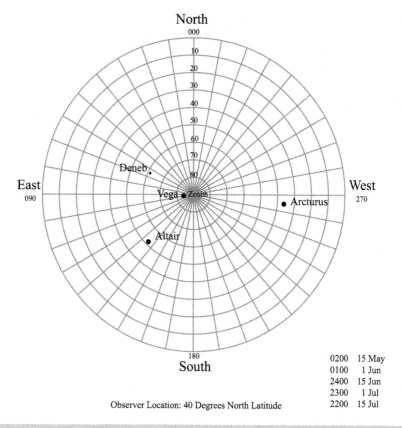

Fig. C.7 North 40 degrees July

Appendix C: Star Charts for Urban Areas with Significant Light Pollution

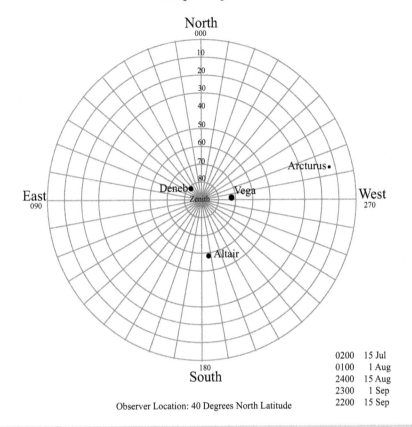

Fig. C.8 North 40 degrees August

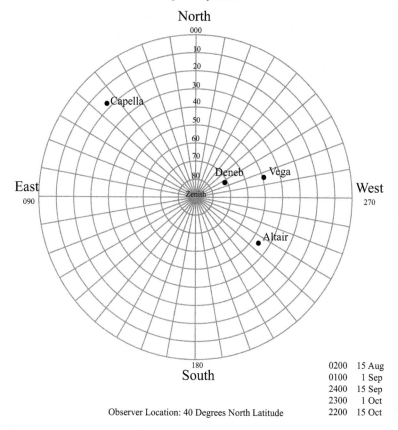

Fig. C.9 North 40 degrees September

Appendix C: Star Charts for Urban Areas with Significant Light Pollution 179

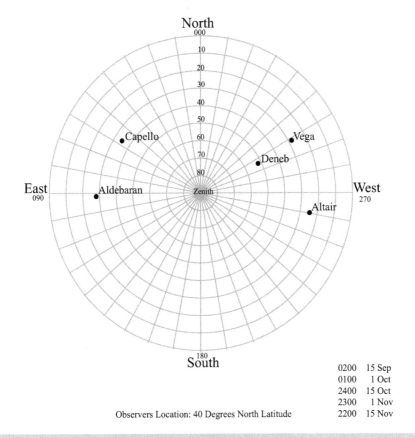

Fig. C.10 North 40 degrees 20 October

Appendix C: Star Charts for Urban Areas with Significant Light Pollution

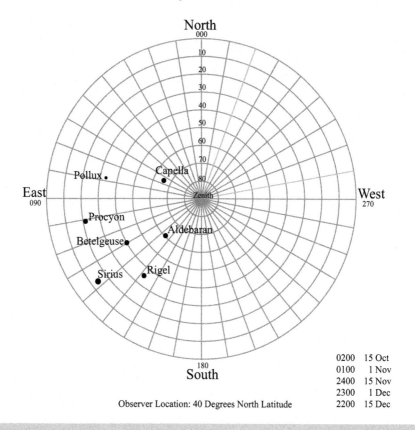

Fig. C.11 North 40 degrees November

Appendix C: Star Charts for Urban Areas with Significant Light Pollution 181

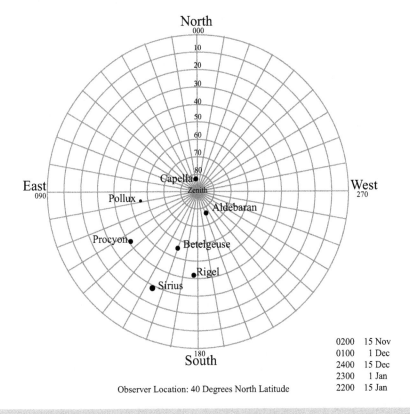

Fig. C.12 North 40 degrees December

Southern Hemisphere at Latitude 30 Degrees

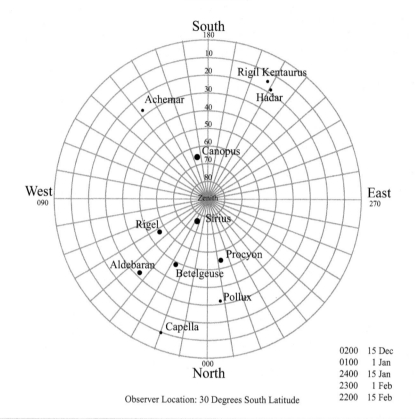

Fig. C.13 South 30 degrees January

Appendix C: Star Charts for Urban Areas with Significant Light Pollution

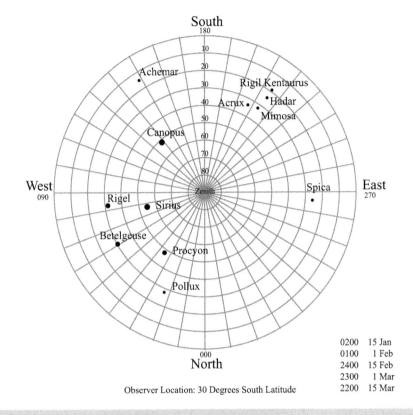

Fig. C.14 South 30 degrees February

184　Appendix C: Star Charts for Urban Areas with Significant Light Pollution

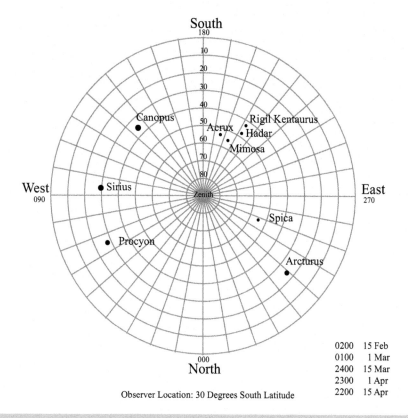

Fig. C.15 South 30 degrees March

Appendix C: Star Charts for Urban Areas with Significant Light Pollution 185

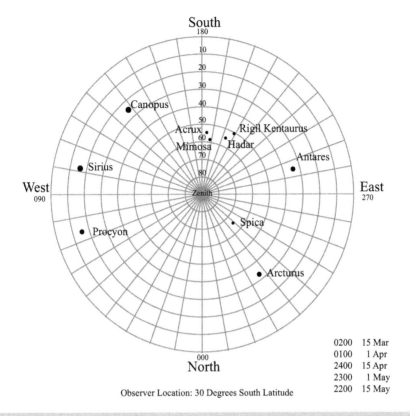

Fig. C.16 South 30 degrees April

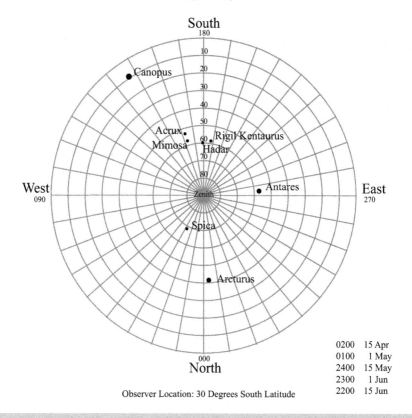

Fig. C.17 South 30 degrees May

Appendix C: Star Charts for Urban Areas with Significant Light Pollution

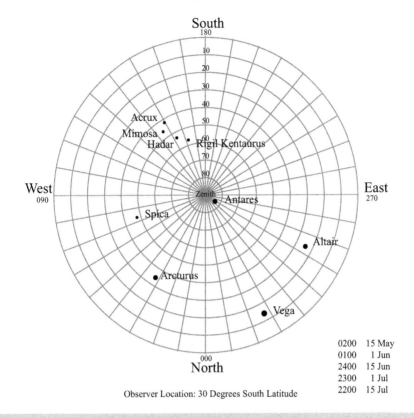

Fig. C.18 South 30 degrees June

188　　Appendix C: Star Charts for Urban Areas with Significant Light Pollution

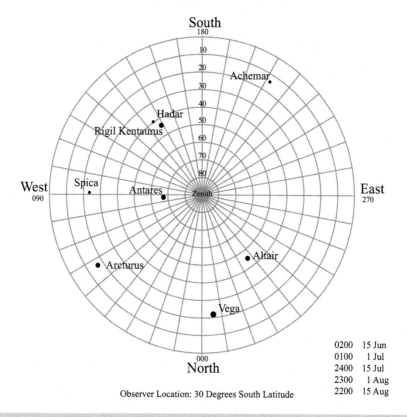

Fig. C.19 South 30 degrees July

Appendix C: Star Charts for Urban Areas with Significant Light Pollution 189

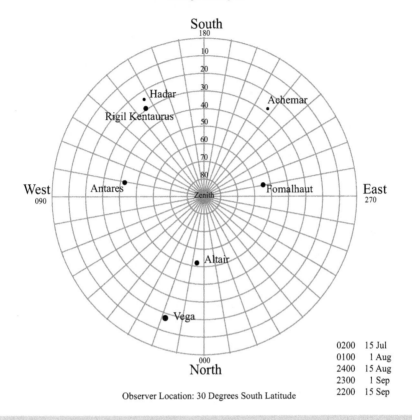

Fig. C.20 South 30 degrees August

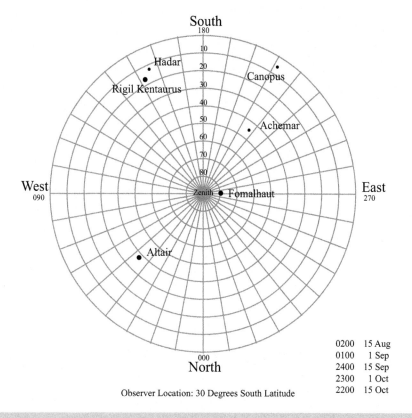

Fig. C.21 South 30 degrees September

Appendix C: Star Charts for Urban Areas with Significant Light Pollution

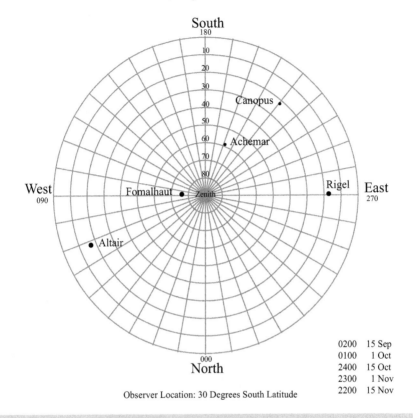

Fig. C.22 South 30 degrees October

192 Appendix C: Star Charts for Urban Areas with Significant Light Pollution

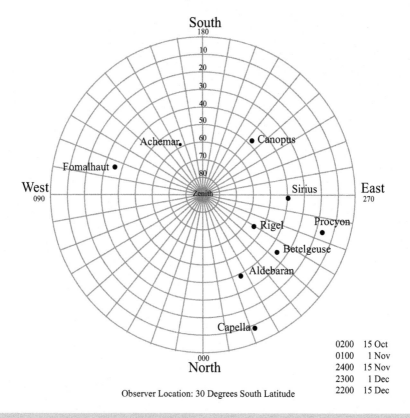

Fig. C.23 South 30 degrees November

Appendix C: Star Charts for Urban Areas with Significant Light Pollution

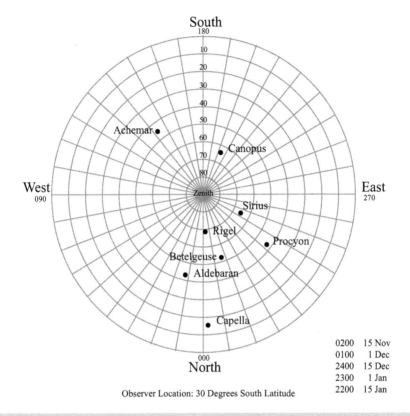

Fig. C.24 South 30 degrees December

Appendix D

Rack and Pinion Focuser Tune-UP

The Synta Short Tube 80 mm f/5 achromatic refractor, sold as either an Orion ST-80 or a SkyWatcher StarTravel 80, is one heck of a bargain given its price range. In the USA the basic OTA starts at around $100. The ST80 sees service as a rich field telescope, a finder scope, and a guide telescope. The telescope is also very suited for usage with an astro-video camera and some use it as a camera telescope for astrophotography.

The telescope comes with an all metal 1.25 inch rack and pinion focuser. The focuser is stiff making achieving a fine focus difficult. One solution is to replace the focuser with a one or two speed Crayford which should transform the telescope into another level of performance but at a price that is greater than the OTA to begin with. This seems to defeat the purpose of buying an inexpensive telescope in the first place. However it is a tribute to the telescope's optics and versatility that people will spend more than the telescope costs to upgrade its focuser.

Another solution, albeit not as good as a Crayford focuser, is to tune-up the stock 1.25 inch rack and pinion metal focuser that comes with the ST-80. The focuser from the factory on a typical ST-80A is stiff and generally has some sticksion. The drawtube wobbles or droops when fully extended. This makes finding the focus point something that is tedious to do. Getting close is easy enough but an exact focus seems more a matter of luck than skill.

Fortunately the focuser can easily be "tuned-up" and provide acceptable performance. The task is not difficult and requires no special tools or skills. Here is what you need to do to enhance the operation of the ST80 focuser:

- Three screws attach the stock rack and pinion focuser to the telescope's tube. Remove these screws and the focuser from the telescope (Fig. D.1)
- Next, remove the four screws on the bottom of the focuser, remove the pinion gear and shaft as well as the draw tube from the focuser (Fig. D.2)

Fig. D.1 Focuser tube screws

Fig. D.2 Pinion gear adjustment screws

- Clean all the Chinese grease from all parts of the focuser.
- Inside the focuser, the draw tube rests on some teflon or nylon strips. Also there is an adjustable drawtube tensioning plate. Put some grease on the teflon strips, the drawtube tensioning plate, and on the rack's gear teeth. **Use grease sparingly as you don't want grease dripping into the interior of your telescope.**

Appendix D: Rack and Pinion Focuser Tune-UP

Fig D.3 Drawtube tensioning adjustment screws

- Reassemble the focuser.
- The bottom of the focuser has four screws (Fig. D.2). These screws adjust the tension on the rack and pinion gearing; too loose and the gears will not mesh (the drawtube can even fall out); too tight and the focuser will be stiff. Adjust these screws until your rack and pinion gearing are fully meshed and stop tightening them when you feel more than a slight resistance to turning the pinion gear. Try to keep the tension equal in each screw. You have to play with this as describing it is difficult. If a small amount of backlash remains in the rack and pinion gear set it is best not try to take the back lash out by tightening the adjustment screws further. Simply make the last focusing adjustment using a slight forward movement of the drawtube toward the telescope.
- If you look on top of the focuser, at either end you will see a very tiny hole (Fig. D.3). This is were the drawtube tensioning adjustment screws are located. You will need a very tiny Allen Wrench (01.5 mm). Slowly adjust each screw to the point where you have no wobble then back-off ever so slightly.
- put the focuser back on the telescope tube.

Index

A
Achromatic refractors, 46, 195
ADC. *See* Analog to digital converter
AGC. *See* Automatic gain control (AGC)
Alignment stars, 47, 51, 71, 98, 130
Alpha Centauri, 73, 74, 89
Alt-azimuth, 3, 4, 6, 17, 23, 38, 46, 48–53, 57, 81, 92, 94–96, 103, 104, 106, 114, 116, 117, 123, 129, 130
Alt-PrintScreen, 118
Anaglyph, 142–145
Analog output, 2, 10, 38, 44, 110
Analog signal, 2, 6, 11, 17, 37, 79, 109, 111, 112, 114
Analog television signal, 1, 37, 57, 111
Analog to digital converter, 1, 37, 94, 111–112, 129, 133, 163
Analog to HDMI converter, 144, 145
Analog video output, vi, vii, 10, 92–94, 118
Angstroms, 77
Aperture, vi, vii, 3, 4, 21, 22, 46–51, 54, 57, 68–71, 75, 76, 84, 85, 90, 91, 94, 95, 97, 108, 127, 129, 130, 155, 158
Apparent brightness, 162
Apparent size, 21, 79, 85, 86, 115
Arc minutes, 19–22, 43, 79, 95–97, 114, 115, 157
Arc second, 62–64, 69, 120, 160, 163

Argon, 62
Armchair astronomers, 4, 57
Artificial sky glow, vi–viii, 3, 4, 56, 60–66, 68, 69, 130, 131, 142, 153, 161
Asteroid Belt, 74, 86
Astro Live, 122
Astro Video Systems (AVS), 16, 25, 38, 42, 47, 95, 101, 146, 152
Astronomical Unit (AU), 73, 74, 156
Astronomy Live, 134, 137
Astrophotography, vii, viii, 2, 3, 6, 13, 16, 17, 23, 38, 47–50, 52, 54, 60, 70, 91, 92, 96, 97, 104, 109, 110, 114–118, 120, 123, 128, 140, 141, 146, 152, 156, 195
Astro-video camera, v, 1, 9, 41, 56, 75, 109, 130, 133, 144, 156, 195
Atmosphere, 56, 60, 78, 81, 83, 161
ATW. *See* Auto tracking white balance (ATW)
AU. *See* Astronomical Unit (AU)
Auroras, 60, 161
Automatic gain control (AGC), 16, 29, 30, 34, 37, 103, 113, 118–120, 124
Auto tracking white balance (ATW), 31
Auto white balance (AWB), 31
AutoStakkert!2, 122, 123, 149
AV System, 34, 35
AVSYSTEM menu, 26, 29–31
AWB. *See* Auto white balance (AWB)

B

Back focus, 47, 156
Backlight compensation (BLC), 30
Bahtinov mask, 81, 100, 117, 122
Balcony, 59, 60
Battery pack, 116
Baudrate, 34, 35
Benard cells, 76
Binoviewers, 145, 146
BLC. *See* Backlight compensation (BLC)
Blooming, 12, 13, 15
BMP, 113, 122, 150
BNC female to RCA male adapter, 43
BNC to RCA adapter, 38, 39, 42
Brightness, 22–24, 29, 31, 34, 37, 57, 62–66, 69, 71, 92, 95, 99, 101–103, 113, 114, 118, 120, 158, 161, 162
Broadband nebula filter, 68
Broadcast/broadcasting, vi, 7, 17, 24, 42–44, 104, 129, 130, 133–140, 142, 147, 149, 151
Broadcasting site, 7, 133–134, 136–139
Button, 25, 26, 29, 101, 136, 137, 139

C

Cable, 1, 4, 6, 9, 38, 39, 41–43, 79, 100, 104, 112, 116, 129, 130, 133, 144, 145, 152
Calcium filters, 77
Camcorder, 43, 82, 84
Camera settings, vii, 4, 24–35, 101, 118–120, 124, 160
Candidate objects, 115
Capture, 1, 6, 21, 23, 31, 33, 36, 42–44, 48, 80–82, 93, 94, 96, 97, 109, 111, 113–116, 118–124, 135–137, 148, 149, 151
Capture Video, 113
CCD. *See* Charge coupled device (CCD)
CCTV. *See* Closed circuit TV (CCTV)
CCTV security cameras, v–vii, 9, 18, 24, 25, 39, 156
Celestial sphere, 156–158
Celestron, 3, 21, 49–51, 53, 54, 96, 99, 100, 104–106, 114, 127, 128, 148
Celestron hand controller extension cable, 104
Charge coupled device (CCD), vii, viii, 2, 10, 18, 24, 25, 36, 38, 39, 42, 47, 56, 70, 79, 82, 91, 92, 98, 109, 110, 114, 124, 141, 152, 156
Checklist, 116
Chromatic aberration, 46, 47, 95, 156
Closed circuit TV (CCTV), v–vii, 9–11, 16, 18, 24, 25, 34, 37–39, 93, 104, 110, 152, 155, 156, 160
CLS filter, 4
CMOS. *See* Complementary metal oxide semiconductor (CMOS)
C-mount, 39, 43, 99, 106, 107
Collimated, 97, 98
COLOR MENU, 26, 27, 31, 32
Comm ID, 34
Complementary metal oxide semiconductor (CMOS), 10, 141
Computer capture, 113
Constellation, 120, 124, 157
Convert the analog signal, 6, 93, 111, 144
Corona, 76, 77
CPU, 148, 150, 151
Crayford focuser, 47, 195
Crowd control, 128

D

Dark frame subtraction, 16, 34, 66, 110, 124, 155
Dark glass eyepiece filter, 76
Dark noise, 12, 16
3D Astronomy, 7, 141, 142
Day and Night menu, 26, 27, 31–32
3D-DNR. *See* Three dimensional digital noise reduction (3D-DNR)
Declination
Deep Sky Stacker, 6, 121–124, 148, 149
Deep space, vi–viii, 2, 3, 5, 9, 15, 16, 21–23, 37, 41, 53, 57, 62, 63, 67, 73, 77, 85, 89–110, 114, 115, 118, 121, 122, 125, 127–130, 138, 143, 153, 157, 160
Deep space object, viii, 3, 5, 15, 16, 21–23, 37, 41, 53, 57, 62, 63, 67, 91–93, 95–97, 101, 108–110, 114, 115, 118, 121, 122, 125, 127, 129, 157, 160
Defective pixel correction (DPC), 16, 17, 34, 37, 66, 103
Denkmeier Optical, 145, 146
Desktop computers, 111, 112, 147, 148
Dew shield, 47, 59
Digital conversion, 37, 94, 111–112
Digital signal, vii, 2, 6, 43, 44, 79, 93, 109–111, 113, 129, 144
Digital single lens reflex (DSLR) camera, v, vii, 1, 2, 10, 11, 16, 19, 30, 36, 47, 56, 70, 91, 92, 94, 96–98, 109, 110, 114, 124, 156
Digital Zoom, 32, 33, 99
3 Dimensional Astro Video System, 146
3 Dimensional Digital Noise Reduction (3D-DNR), 14, 16, 17, 34, 36–38, 48, 49, 51, 65, 66, 70, 91, 93, 94, 103, 113, 118–121, 123, 124, 155

Dobson, 19
DPC. *See* Defective pixel correction (DPC)
DSLR. *See* Digital single lens reflex (DSLR) camera
DSO, 16, 38, 91, 157
DSO-1 camera, 25, 38, 42, 95
2D to 3D converter, 144, 145
DVD player, vi, vii, 1, 43, 92, 163
DVD recorder, 43
3D Video astronomy, 143, 145, 146
Dwarf planet, 74, 86
Dynamic range, 30
Dysnomia, 74

E
Effect menu, 26, 27, 32–33
Effective aperture, vii, 54, 90, 127
Electric power availability, 128
Electrical power, 6, 30, 100
ENHANCE, 25, 101, 103
Entry-level astro-video cameras, 6, 9–11, 13–19, 25, 35, 37–39, 42, 43, 51, 54, 65, 66, 94, 110, 114, 115
Equatorial, 3, 4, 6, 17, 23, 34, 38, 46, 48, 49, 52–54, 66, 70, 81, 82, 92, 94–96, 98, 103, 104, 106, 116, 117, 119, 120, 124, 128, 129, 155, 157, 159, 161
Equatorial mount, 3, 4, 6, 23, 34, 38, 46, 48, 49, 52, 53, 66, 70, 81, 82, 92, 94–96, 98, 103, 104, 116, 117, 128, 129, 155, 157, 159, 161
Eris, 74
ETX, 50, 51, 53, 105, 127
Europe, viii, 53, 55, 59, 127
Event objectives, 128
EX VIEW HAD II CCD sensor, 24–25
Expansion, 128, 141
Extension cable, 100, 104
External USB device, 111
Eyeglass, 142–145
Eyepiece, vi, 1, 39, 41, 57, 75, 89, 117, 125, 145

F
FastStar, 3
Field of view (FOV), 2, 3, 6, 18–22, 42, 43, 46–48, 50, 51, 54, 80, 85, 92, 94–99, 108, 114–117, 123, 124, 128, 129, 146, 157, 158
Field rotation, 4, 6, 17, 23, 34, 38, 48–52, 81, 92–94, 96, 103, 104, 114, 116, 122, 123, 158, 165–169
Filaments, 76

Filters, 4, 10, 46, 56, 75, 99, 140, 142
Finderscope, 47
Flaring, 12, 13, 32, 46, 48, 101, 102
Florescent, 61
Flux, 63, 64, 66, 69
Focal lengths, 2, 19, 43, 79, 92, 114, 129, 146, 157, 158
0.5 Focal reducer, 53
0.63 Focal reducer, 24
Focal reducer, 2, 3, 6, 21–23, 39, 41, 46–48, 50, 51, 53, 54, 96–99, 106–108, 114, 115, 117, 124, 130, 135, 161
Focuser, 2, 47, 48, 54, 59, 96, 99, 117, 156, 160, 161, 195–197
Focusing, 54, 60, 78, 81, 99, 117, 122, 123, 125, 128, 146, 197
Formal presentation, 126, 129
FOV. *See* Field of view (FOV)
Frame, 6, 11–14, 16, 17, 23, 34, 37, 42–44, 51, 57, 58, 66, 78, 79, 81–83, 93, 97, 101, 103, 110–114, 118–124, 133, 136, 142, 144, 149, 151, 155, 156
Frame grabber, 6, 17, 37, 42–44, 57, 79, 93, 111–114, 118, 133, 136, 144, 150, 151
Freeze, 32
Future, vii, 1, 41, 56, 87, 113, 137, 141, 142, 146, 152, 153

G
Galaxies, vii, 4, 60, 62, 63, 67, 69, 71, 89, 127, 129, 138, 143, 146, 153, 157–159, 161, 162
Galileo, 89, 161
Gamma, 34, 37, 101, 103, 118
Gegenschein, 60
GEM. *See* German equatorial mount (GEM)
General notes, 37–38
German equatorial mount (GEM), 3, 4, 17, 46, 48, 52, 92, 95, 96, 103, 104, 116, 117, 128, 129, 157
Glare, 57, 60, 66, 83
Globular clusters, 67, 71, 138, 157, 159
Globular star clusters, 128, 158
Google Hangouts, 134, 137
GOTO mount, 2–4, 17, 22, 46–48, 53, 92, 94, 96, 100, 117, 127–129, 133, 158
Green glow, 60
GStar4Capture, 121–123, 149

H
Halogen, 62
Halos, 33, 57, 158

Hand controller, 4, 18, 29, 30, 41–43, 79, 100, 101, 103, 104
Haumea, 74
HDMI, 144, 145, 149–152
Herschel wedge, 75
High definition TV, 93, 145
High pressure sodium, 62
High speed internet access, 130, 133
Highlight (HLI), 30
Human eye, 1, 2, 51, 61–63, 77
Hydrogen beta, 68, 70
Hydrogen-alpha (H-Alpha) filters, 68, 70, 76–78
HyperStar, 3

I
Image brightness, 22–24, 34, 92, 95, 101–103, 158
Image processing, 2, 110
Image size, 2, 3, 6, 19, 22, 43, 46, 48, 53, 85, 92, 94, 99, 114, 158, 159
Imaging, 6, 9, 12, 16, 17, 24, 32, 33, 42, 46, 60, 69, 77, 78, 80, 81, 85, 87, 94, 109–124, 127, 136, 151, 152
Imaging session, 85, 114, 116, 117
Incandescent, 61
Infrared, vi, 9, 46, 48, 68, 103
Infrared cut filter, 48
Installing a frame grabber, 112
INTC setting, 30
Integrated exposure, 4, 6, 10, 13–17, 30, 36, 65, 93, 110
Integrated exposure time, 4, 6, 10, 13–17, 30, 36, 65, 93
Integrating/integration, 4, 11, 13–17, 22–24, 30, 35–38, 42, 48–51, 65, 66, 70, 71, 82, 91–93, 96, 101–104, 110, 114, 117–121, 123, 124, 140, 141, 146
Integrating camera, 4, 11, 13, 14, 17, 35–37, 93, 101, 110, 141, 146
Intel, 112, 130, 147, 150
Interlace/interlacing, 11, 37, 111, 157
Internal cooling, 18, 42, 103
International Occultation and Timing Association (IOTA), 84, 87
Internet, v, vii, 1, 7, 11, 37, 43, 44, 46, 47, 75, 79, 97, 100, 104, 124, 127, 128, 130, 133, 134, 138, 139, 141–144, 147–150, 163
INTMUL, 16, 34, 37, 38, 103, 118, 119
INTMUL (3D-DNR), 16, 34, 37, 38, 118
iOptron, 48–52, 54, 76, 77, 96, 114, 127, 128

IOTA. *See* International Occultation and Timing Association (IOTA)
IRIS, 18, 30
ISO settings, 30

J
JPG, 113, 150
Jupiter, 74, 80, 84–86, 125, 128

K
Kuiper Belt, 74, 162

L
Language, 34, 35, 122, 139
Laptop computer, 4, 7, 111, 112, 133, 141, 147–149, 151, 152
Laser, 129
Law enforcement, 129
LED, 31, 32, 62, 142, 152
Light pollution, 2, 55–71, 80, 128, 152, 153, 162, 169–195
Light pollution reduction (LPR) filters, 67–71
Light screen, 56, 58–60
Light shield, 58–60
Light year (ly), 73, 74, 89, 95, 120, 124, 159, 160
Lightweight mounts, 23, 46, 49, 57, 127, 129
Line filter, 68, 70, 71
Linear stack, 12–14, 36, 162
Live broadcast, 134, 138, 139
Live stack, 122, 123, 151
Live video broadcast/broadcasting, 130, 133–140, 151
Live view, 67, 113, 118, 123, 163
LPR. *See* Light pollution reduction (LPR) filters
Low light, v, vi, 9, 12, 16, 92, 146, 152, 155, 156
Low pressure sodium, 62, 68
Low resolution, 94, 109, 110
Lunar Impact Monitoring, 82–83
Lunar Observation, 80–81
Lunar photography, 81, 82
LunarScan software, 82

M
M31, 115
M36, 119, 120
M42, 21, 101, 102
M45, 115, 127, 160

Magnitude, 4, 17, 56, 57, 61–64, 66, 69, 71, 90, 98, 114, 120, 124, 161, 162
Makemake, 74
Maksutov, 23, 50, 53, 54, 105–108, 156
Maksutov Cassegrain Telescopes, 23, 50, 53, 54, 105–108, 156
MallinCam, 39, 42, 95, 101, 130, 152
Mars, 74, 85, 86
Meade, 3, 21, 47, 49–51, 53, 76, 77, 105, 114, 127
Megapixels, 10
Mercury, 62, 68, 74, 85, 86, 152
Mercury vapor, 62, 68
Meridian, 4, 116
Messier, C., 115, 159
Messier object, 16, 115, 159
Mirror, 22, 32, 59, 95, 100, 110, 116, 155, 156, 158, 160
Miscellaneous issues such as air traffic, 128
Monitor, vi, vii, 1, 2, 4, 25, 31, 32, 34, 39, 41–43, 50, 57, 59, 60, 78, 79, 82–84, 92, 93, 96, 99–102, 104, 113, 117, 118, 121, 126–131, 136, 143–145, 147, 151, 160, 163
Moon, vi, 5, 9, 19, 21, 30, 60–62, 69, 74, 79–86, 97, 125, 128, 138, 157, 162
Motion menu, 26, 28, 33
Mumbler, 127

N

Narrowband nebula filter, 68
Natural sky glow, 56, 60–62, 68
Near real-time, 60, 71, 91, 96, 100, 103, 113, 117, 118, 140, 142–146
Near Real-Time Viewing, 60, 117, 118, 140, 146
Nebula, 21, 62–69, 71, 96, 101, 102, 120, 159, 161
Nebula filter, 67, 68, 71
Nebula line filter, 68
Nebulae, v, vii, 4, 7, 21, 62, 68, 69, 71, 77, 89, 126–128, 138, 143, 146, 155, 157, 159, 161, 162
Neon, 62
Neptune, 74, 85
Newton, I., 160
Newtonian, 2, 19, 23, 46, 47, 50, 59, 96, 99, 156, 160
Night Skies Network, 7, 134, 136–137
Noise, viii, 6, 10, 42, 63, 92, 110, 155
Nose piece, 18
NTSC Television Standard, 6, 10, 11

O

Observation session, 22, 59, 95, 98
Occultations, 83–84, 87, 123
On screen display (OSD), 18, 25, 26, 29, 35, 38, 42, 94, 101, 103, 114, 160
Oort Cloud, 73, 162
Open cluster, 124, 157, 159, 160
Opik Cloud, 73
Orion, 21, 48–54, 96, 100, 101, 105, 114, 127, 195
OSD. *See* On screen display (OSD)
OTG cables, 116
Outreach, vii, 5–6, 17, 79, 86, 125–131
Outreach activities, 5, 125, 127
Outreach event, 6, 126–130
Oxygen III, 68, 70

P

PAL Television Standard, 1, 6, 10, 91, 156
Parafocal, 99, 100, 103, 117
Parafocal eyepiece, 99, 103, 117
Parafocal ring, 99, 100, 117
Parsec(pc), 160
Payload, 48, 49, 52, 129, 161
Payload capacity, 49, 129
PCI card, 111
Photo processing program, 113, 114, 119, 121
Photographs, 12, 55, 57, 67, 71, 77, 78, 85, 89, 96, 109, 111, 115, 116, 139, 142–143
Photon noise, 12, 16, 63–65, 70
Photons, 10, 12, 15, 16, 22, 23, 48, 62–67, 69, 70, 101, 159, 161
Pier, 92
Pixel, 10–13, 15–17, 34, 37, 48, 63–66, 70, 93, 94, 101, 102, 109, 110, 162
Pleiades, 115, 160
Pluto, 74, 85, 86
Poisson distribution, 12, 13
Polar aligned/alignment, 4, 17, 23, 33, 47, 48, 53, 66, 98, 104
Portable observatory, 57
Predict, 141–142
Procamp menu, 28, 34
Prominence, 76, 77
Public facilities, 128
Public speaker, 128
Public transportation, viii, 57, 59
Push, 31, 139

Q

Quick capture, 122

R

Rack and Pinion Focuser, 195–197
Rated payload, 161
Read noise, 12, 16
Real-time, v, vi, 2, 17, 41, 60, 63, 71, 78, 79, 81, 91, 96, 100, 103, 113, 114, 117, 118, 136, 140, 142–146
Real-time conversion, 143
Refractor, 2, 19, 22, 23, 46–48, 50–54, 57, 63, 77, 89–91, 94–97, 99, 102, 103, 114, 115, 130, 138, 160, 161, 195
Registax, 121–123, 149
Remote hand controller, 30, 43, 101
Remote viewing, 2, 5, 100, 142
Resolution, 10, 43, 90–92, 94, 109, 110, 118, 145, 150, 152
Reticle eyepiece, 47, 98, 123
Revolution Imager, 39, 42, 95, 101
Rules, 3, 46, 90, 136, 138–140, 147, 160

S

Saturn, 74, 80, 85, 86, 125, 128
Schmidt Cassegrain telescope (SCT), 3, 19, 21–23, 32, 46, 47, 49, 50, 90, 94, 95, 97, 105–108, 115, 129, 138, 156
Screen capture, 109, 113, 118, 120, 135
Screen capture programs, 114, 119
Screen save, 113
SCTs. *See* Schmidt Cassegrain telescope (SCT)
Security camera, v–vii, 9, 18, 24, 25, 39, 156
4 SE mount, 53
5 SE mount, 53
Sense up, 29, 30
Sensors, 2, 10, 42, 64, 79, 92, 109, 151, 159
4SE telescope, 106, 108
Setup for astrophotography, 114
SharpCap, 114, 122, 148, 149
Sharpness, 32, 33, 48, 52, 79, 118
Short tube refractor, 2, 19, 47, 48, 53, 54, 96, 115
Shutter speed, 6, 10, 30, 93, 99
Signal, vi, 1, 10, 43, 57, 79, 93, 109, 129, 144
Signal to noise ratio (SNR), 10, 13–15, 22, 23, 64–66, 69–71, 101, 110, 113, 121
Sky glow, vi–viii, 56, 57, 60–71, 130, 142, 152, 153, 161
SkyWatcher, 48–51, 53, 96, 100, 105, 114, 127, 195
SLT, 50, 51, 53, 54, 96, 98, 127
SmartEQ PRO, 48, 49, 52, 54, 96
Snapshot, 39, 113, 122

SNR. *See* Signal to noise ratio (SNR)
Software, 42, 44, 45, 82, 98, 109, 135, 136, 142, 143, 150
Solar flare, 74, 76, 77
Solar System, 73–87, 123, 127, 162
Sony EX VIEW, 24
Square root, 12–16, 63–65
ST-80A, 195
Stabilizer, 32, 33, 118
Stacking, 6, 11–14, 16, 17, 21, 36, 48, 52, 65, 66, 71, 81, 93, 101, 109, 110, 113, 114, 120–124, 149, 151, 155, 162
Stack real-time, 114
Stars, v, 3, 17, 46, 57, 73, 89, 116, 126, 138, 146, 156
Star cluster, 71, 119, 127, 128, 146, 158, 160, 161
Starlight, 9, 10, 141, 156
StarNavigator, 50
Start Capture, 113
StarTravel, 53, 96, 195
Statistical stack, 12–14, 16, 36, 162
Stellar magnitude, 162
Stray light, 57–60, 66, 128
Stray light from TV monitors, 128
Street corner astronomer, 128
Summer, vi, viii, 4, 57, 76, 79, 103, 104, 138
Sun, v, 30, 60, 61, 73–83, 85, 86, 89, 156, 157, 162
Sunspots, 76, 77
Surface brightness, 57, 62–64, 66, 69, 120, 162
Sync, 35, 98
SynScan, 50, 51, 53, 96, 100, 127
System menu, 29, 34, 35

T

Tablet, 7, 42, 44, 45, 93, 105, 112, 116, 129–131, 133, 141, 146–152
Telescope, v, 1, 16, 41, 55, 73, 89, 109, 125, 133, 142, 155, 156
Telescope operator skills, 128
Test Bar menu, 28, 33
Thermal noise, 103
Three-dimensional binoculars, 146
TIFF, 122
Title, 25, 34, 35, 137
Tracking movements, 50–52, 94, 108
Trespass light, 56–60, 152
Tripod, 3, 18, 23, 41, 50, 53, 92, 98, 100, 129, 133, 161
Tripod vibration, 3, 50, 53
Tune-up, 96, 195–197

Index

TV monitor, vii, 2, 41–43, 50, 59, 60, 78, 79, 92, 93, 96, 100–102, 104, 126–131, 144, 145, 147, 151
TV set, 11, 37, 41, 43, 92, 126, 144, 145, 157

U

Ultra-high contrast (UHC) filter, 68
Uranus, 74, 85, 86
Urban, viii, 3, 55, 56, 69, 71, 80, 130–131, 142, 152–153, 169–195
Urban astronomers, 56
Urban astronomy, 152–153
Urbanization, 55, 56, 141
USB cable, 38, 43, 116
USB port, 6, 45, 148–151
USB-A OTG, 149

V

Variable reducers, 21, 47, 97
VCR, 43, 92
Velcro, 100
Venus, 74, 85, 86
Vibration, 3, 17, 23, 49–51, 53, 59, 60, 95, 96, 108
Video astronomy, v, 1, 9, 43, 57, 83, 90, 111, 125, 134, 141, 163
Video Astronomy Live, 134–136
Video capture device, 6, 42–44, 93, 94, 111, 136
Video capture program, 113, 119, 121–124

Video coax RG 59, 43
Video editing, 113, 118
Video telescope, vi, 2, 17, 41, 56, 75–80, 89, 112, 125, 134, 144
Vignetting Correction, 34
Visual astronomy, v, 158
Visual observing, 60, 94

W

WDR. *See* Wide dynamic range (WDR)
Weather, 42, 76, 135, 138
White light, 61, 75, 77, 79
White light filters, 77
Wide dynamic range (WDR), 30
WiFi AV, 130
Windows 8, 112
Windows 10, 44, 112, 130, 146–152
Windows 10 tablet (Win10 tablet), 42, 44, 112, 130, 131, 146–152
Wire management, 100
World Health Organization, 55

X

Xenon, 62
XSplit Broadcaster, 135, 148, 149

Z

Zenith, 4, 6, 49, 116, 155, 158
Zodiacal light, 60, 61
Zoom, 32, 33, 99, 118, 123

Printed by Printforce, the Netherlands